METHODS IN MOLECULAR BIOLOGY

For further volumes:
http://www.springer.com/series/7651

For over 35 years, biological scientists have come to rely on the research protocols and methodologies in the critically acclaimed *Methods in Molecular Biology* series. The series was the first to introduce the step-by-step protocols approach that has become the standard in all biomedical protocol publishing. Each protocol is provided in readily-reproducible step-by-step fashion, opening with an introductory overview, a list of the materials and reagents needed to complete the experiment, and followed by a detailed procedure that is supported with a helpful notes section offering tips and tricks of the trade as well as troubleshooting advice. These hallmark features were introduced by series editor Dr. John Walker and constitute the key ingredient in each and every volume of the *Methods in Molecular Biology* series. Tested and trusted, comprehensive and reliable, all protocols from the series are indexed in PubMed.

Bioluminescent Imaging

Methods and Protocols

Edited by

Steven Ripp

The Center for Environmental Biotechnology, The University of Tennessee, Knoxville, TN, USA

490 BioTech Inc., Knoxville, TN, USA

Humana Press

Editor
Steven Ripp
The Center for Environmental Biotechnology
The University of Tennessee
Knoxville, TN, USA

490 BioTech Inc.
Knoxville, TN, USA

ISSN 1064-3745 ISSN 1940-6029 (electronic)
Methods in Molecular Biology
ISBN 978-1-4939-9939-2 ISBN 978-1-4939-9940-8 (eBook)
https://doi.org/10.1007/978-1-4939-9940-8

This Humana imprint is published by the registered company Springer Science+Business Media, LLC, part of Springer Nature.
The registered company address is: 233 Spring Street, New York, NY 10013, U.S.A.

Preface

Bioluminescent imaging uses the power of light to visualize, monitor, measure, and track biological processes as they occur in vitro in the laboratory test tube, in vivo in the live animal, ex vivo in animal tissues, or in planta in intact plants. This imaging modality has seen widespread application to address basic research questions in cellular and molecular biology and has evolved as a reporter gene toolset now commonly used in biomedical fields (pathogenesis, oncology, immunology, neurology) and for environmental biomonitoring. The chapters in this volume provide detailed protocols and methodologies required to perform bioluminescent imaging at all of its multifaceted stages to enable the reader to integrate this technology into their laboratory-based imaging experiments.

Knoxville, TN, USA *Steven Ripp*

Contents

Preface . . . . *v*
Contributors . . . . *ix*

Part I In Vitro/Ex Vivo Imaging Applications

1 High-Throughput Bioluminescence Imaging and Reporter Gene Assay with 3D Spheroids from Human Cell Lines 3
Maria Maddalena Calabretta, Laura Montali, Antonia Lopreside, Elisa Michelini, and Aldo Roda

2 Bioluminescence Resonance Energy Transfer (BRET) Coupled Near-Infrared Imaging of Apoptotic Cells 15
Setsuko Tsuboi and Takashi Jin

3 High-Throughput Analysis of Endocrine-Disrupting Compounds Using BLYES and BLYAS Bioluminescent Yeast Bioassays 29
Tingting Xu, Anna Young, Jasleen Narula, Gary Sayler, and Steven Ripp

4 Bioluminescent Imaging of Single Bacterial Cells Using an Enhanced *ilux* Operon 43
Carola Gregor

5 Bioluminescent Imaging and Tracking of Bacterial Transport in Soils 53
Jie Zhuang, Weipeng Liu, Liqiong Yang, Jia Kang, and Xiaoming Zhang

Part II In Vivo Imaging of Microbial Pathogenesis

6 In Vivo Bioluminescent Imaging of *Yersinia ruckeri* Pathogenesis in Fish 69
José A. Guijarro and Jessica Méndez

7 In Vivo Bioluminescent Monitoring of Parasites in BALB/c Mouse Models of Cutaneous Leishmaniasis Drug Discovery 81
Diana Caridha, Susan Leed, and Alicia Cawlfield

8 Multiplex Imaging of Polymicrobial Communities—Murine Models to Study Oral Microbiome Interactions 107
Jens Kreth, Yasser M. Abdelrahman, and Justin Merritt

9 Bioluminescence Imaging to Study Mature Biofilm Formation by *Candida* spp. and Antifungal Activity In Vitro and In Vivo 127
Katrien Van Dyck, Patrick Van Dijck, and Greetje Vande Velde

Part III In Vivo Small Animal/Organism Imaging Applications

10 Bioluminescence Imaging of Neuroinflammation in a Mouse Model of Parkinson's Disease 147
Maria Eugenia Bernis and Gültekin Tamgüney

11 Continual Conscious Bioluminescent Imaging in Freely Moving Mice 161
Juan Antinao Diaz, Amy Geard, Lorna M. FitzPatrick, Juliette M. K. M. Delhove, Suzanne M. K. Buckley, Simon N. Waddington, Tristan R. McKay, and Rajvinder Karda

12 In Vivo Bioluminescent Imaging of Marburg Virus in a Rodent Model 177
Shan Lei, Weijin Huang, Youchun Wang, and Qiang Liu

13 Continuous and Real-Time In Vivo Autobioluminescent Imaging in a Mouse Model 191
Derek Yip, Andrew Kirkpatrick, Tingting Xu, Tom Masi, Stacy Stephenson, Steven Ripp, and Dan Close

14 In Vivo Tracking of Tumor-Derived Bioluminescent Extracellular Vesicles in Mice 203
Prakash Gangadaran and Byeong-Cheol Ahn

15 Bioluminescence Imaging in the Chick Chorioallantoic Membrane Assay 211
Benedict Jefferies, Zhichao Tong, and Roman Nawroth

16 Application of a cybLuc Aminoluciferin for Deep Tissue Bioluminescence Imaging in Rodent Models 219
Xiang Li and Minyong Li

Part IV In Planta Imaging Applications

17 Application of Single-Cell Bioluminescent Imaging to Monitor Circadian Rhythms of Individual Plant Cells 231
Tomoaki Muranaka and Tokitaka Oyama

Index *243*

Contributors

YASSER M. ABDELRAHMAN • *Department of Restorative Dentistry, School of Dentistry, Oregon Health and Science University, Portland, OR, USA; Department of Microbiology and Immunology, Faculty of Pharmacy, Cairo University, Cairo, Egypt*

BYEONG-CHEOL AHN • *Department of Nuclear Medicine, School of Medicine, Kyungpook National University, Daegu, Republic of Korea; Department of Nuclear Medicine, Kyungpook National University Hospital, Daegu, Republic of Korea*

MARIA EUGENIA BERNIS • *German Center for Neurodegenerative Diseases (DZNE), Bonn, Germany*

SUZANNE M. K. BUCKLEY • *Gene Transfer Technology Group, Institute for Women's Health, University College London, London, UK*

MARIA MADDALENA CALABRETTA • *Department of Chemistry "Giacomo Ciamician", University of Bologna, Bologna, Italy*

DIANA CARIDHA • *Walter Reed Army Institute of Research, Silver Spring, MD, USA*

ALICIA CAWLFIELD • *Dwight D. Eisenhower Army Medical Center, Fort Gordon, GA, USA*

DAN CLOSE • *490 BioTech Inc., Knoxville, TN, USA*

JULIETTE M. K. M. DELHOVE • *Robinson Research Institute, University of Adelaide, Adelaide, SA, Australia*

JUAN ANTINAO DIAZ • *Gene Transfer Technology Group, Institute for Women's Health, University College London, London, UK*

LORNA M. FITZPATRICK • *Centre for Biomedicine, Manchester Metropolitan University, Manchester, UK*

PRAKASH GANGADARAN • *Department of Nuclear Medicine, School of Medicine, Kyungpook National University, Daegu, Republic of Korea; Department of Nuclear Medicine, Kyungpook National University Hospital, Daegu, Republic of Korea*

AMY GEARD • *UCL School of Pharmacy, University College London, London, UK; Wits/SAMRC Antiviral Gene Therapy Research Unit, Faculty of Health Sciences, University of the Witwatersrand, Johannesburg, South Africa*

CAROLA GREGOR • *Department of NanoBiophotonics, Max Planck Institute for Biophysical Chemistry, Göttingen, Germany*

JOSÉ A. GUIJARRO • *Departamento de Biología Funcional, Facultad de Medicina, IUBA, Universidad de Oviedo, Oviedo, Asturias, Spain*

WEIJIN HUANG • *Division of HIV/AIDS and Sex-Transmitted Virus Vaccines, National Institutes for Food and Drug Control, Beijing, China*

BENEDICT JEFFERIES • *Department of Urology, Klinikum rechts der Isar, Technical University of Munich, Munich, Germany*

TAKASHI JIN • *Riken Center for Biosystems Dynamics Research, Riken, Osaka, Japan*

JIA KANG • *Key Laboratory of Pollution Ecology and Environmental Engineering, Institute of Applied Ecology, Chinese Academy of Sciences, Shenyang, China; The University of Chinese Academy of Sciences, Beijing, China*

RAJVINDER KARDA • *Gene Transfer Technology Group, Institute for Women's Health, University College London, London, UK*

ANDREW KIRKPATRICK • *490 BioTech Inc., Knoxville, TN, USA*

JENS KRETH • *Department of Restorative Dentistry, School of Dentistry, Oregon Health and Science University, Portland, OR, USA*
SUSAN LEED • *Walter Reed Army Institute of Research, Silver Spring, MD, USA*
SHAN LEI • *Division of HIV/AIDS and Sex-Transmitted Virus Vaccines, National Institutes for Food and Drug Control, Beijing, China*
MINYONG LI • *Department of Medicinal Chemistry, Key Laboratory of Chemical Biology of Natural Products (MOE), School of Pharmacy, Shandong University, Jinan, Shandong, China*
XIANG LI • *Department of Medicinal Chemistry, Key Laboratory of Chemical Biology of Natural Products (MOE), School of Pharmacy, Shandong University, Jinan, Shandong, China*
QIANG LIU • *Division of HIV/AIDS and Sex-Transmitted Virus Vaccines, National Institutes for Food and Drug Control, Beijing, China*
WEIPENG LIU • *Key Laboratory of Pollution Ecology and Environmental Engineering, Institute of Applied Ecology, Chinese Academy of Sciences, Shenyang, China; The University of Chinese Academy of Sciences, Beijing, China*
ANTONIA LOPRESIDE • *Department of Chemistry "Giacomo Ciamician", University of Bologna, Bologna, Italy*
TOM MASI • *Department of Surgery, Graduate School of Medicine, The University of Tennessee Medical Center, Knoxville, TN, USA*
TRISTAN R. MCKAY • *Centre for Biomedicine, Manchester Metropolitan University, Manchester, UK*
JESSICA MÉNDEZ • *Departamento de Biología Funcional, Facultad de Medicina, IUBA, Universidad de Oviedo, Oviedo, Asturias, Spain*
JUSTIN MERRITT • *Department of Restorative Dentistry, School of Dentistry, Oregon Health and Science University, Portland, OR, USA; Department of Molecular Microbiology and Immunology, Oregon Health and Science University, Portland, OR, USA*
ELISA MICHELINI • *Department of Chemistry "Giacomo Ciamician", University of Bologna, Bologna, Italy; INBB, Istituto Nazionale di Biostrutture e Biosistemi, Rome, Italy; Health Sciences and Technologies-Interdepartmental Center for Industrial Research (HST-ICIR), University of Bologna, Bologna, Italy*
LAURA MONTALI • *Department of Chemistry "Giacomo Ciamician", University of Bologna, Bologna, Italy*
TOMOAKI MURANAKA • *Center for Ecological Research, Kyoto University, Otsu, Shiga, Japan*
JASLEEN NARULA • *The Center for Environmental Biotechnology, The University of Tennessee, Knoxville, TN, USA*
ROMAN NAWROTH • *Department of Urology, Klinikum rechts der Isar, Technical University of Munich, Munich, Germany*
TOKITAKA OYAMA • *Department of Botany, Graduate School of Science, Kyoto University, Kyoto, Japan*
STEVEN RIPP • *The Center for Environmental Biotechnology, The University of Tennessee, Knoxville, TN, USA; 490 BioTech Inc., Knoxville, TN, USA*
ALDO RODA • *Department of Chemistry "Giacomo Ciamician", University of Bologna, Bologna, Italy; INBB, Istituto Nazionale di Biostrutture e Biosistemi, Rome, Italy*
GARY SAYLER • *The Center for Environmental Biotechnology, The University of Tennessee, Knoxville, TN, USA; 490 BioTech Inc., Knoxville, TN, USA*
STACY STEPHENSON • *Department of Surgery, Graduate School of Medicine, The University of Tennessee Medical Center, Knoxville, TN, USA*

GÜLTEKIN TAMGÜNEY • *Institute of Complex Systems — Structural Biochemistry (ICS-6), Forschungszentrum Jülich GmbH, Jülich, Germany; Institut für Physikalische Biologie, Heinrich-Heine-Universität Düsseldorf, Düsseldorf, Germany*
ZHICHAO TONG • *Department of Urology, Klinikum rechts der Isar, Technical University of Munich, Munich, Germany*
SETSUKO TSUBOI • *Riken Center for Biosystems Dynamics Research, Riken, Osaka, Japan*
PATRICK VAN DIJCK • *VIB-KU Leuven Center for Microbiology, Leuven, Belgium; Laboratory of Molecular Cell Biology, Institute of Botany and Microbiology, KU Leuven, Leuven, Belgium*
KATRIEN VAN DYCK • *VIB-KU Leuven Center for Microbiology, Leuven, Belgium; Laboratory of Molecular Cell Biology, Institute of Botany and Microbiology, KU Leuven, Leuven, Belgium*
GREETJE VANDE VELDE • *Biomedical MRI/MoSAIC, Department of Imaging and Pathology, KU Leuven, Leuven, Belgium*
SIMON N. WADDINGTON • *Gene Transfer Technology Group, Institute for Women's Health, University College London, London, UK; Wits/SAMRC Antiviral Gene Therapy Research Unit, Faculty of Health Sciences, University of the Witwatersrand, Johannesburg, South Africa*
YOUCHUN WANG • *Division of HIV/AIDS and Sex-Transmitted Virus Vaccines, National Institutes for Food and Drug Control, Beijing, China*
TINGTING XU • *The Center for Environmental Biotechnology, The University of Tennessee, Knoxville, TN, USA*
LIQIONG YANG • *Key Laboratory of Pollution Ecology and Environmental Engineering, Institute of Applied Ecology, Chinese Academy of Sciences, Shenyang, China; The University of Chinese Academy of Sciences, Beijing, China*
DEREK YIP • *490 BioTech Inc., Knoxville, TN, USA*
ANNA YOUNG • *The Center for Environmental Biotechnology, The University of Tennessee, Knoxville, TN, USA*
XIAOMING ZHANG • *College of Desert Control Science, Inner Mongolia Agricultural University, Hohhot, China*
JIE ZHUANG • *Department of Biosystems Engineering and Soil Science, Center for Environmental Biotechnology, The University of Tennessee, Knoxville, TN, USA; Key Laboratory of Pollution Ecology and Environmental Engineering, Institute of Applied Ecology, Chinese Academy of Sciences, Shenyang, China*

Part I

In Vitro/Ex Vivo Imaging Applications

Chapter 1

High-Throughput Bioluminescence Imaging and Reporter Gene Assay with 3D Spheroids from Human Cell Lines

Maria Maddalena Calabretta, Laura Montali, Antonia Lopreside, Elisa Michelini, and Aldo Roda

Abstract

3D cell culture models represent an attractive approach to decode intracellular and intercellular signaling, providing biologically relevant information and predictive data. Bioluminescent reporter gene assays and bioluminescence imaging in 3D cell models are very promising bioanalytical tools for several applications.

Here we report a very straightforward method for bioluminescence imaging and bioluminescent reporter gene assays in 3D cell-culture models. Both the assays can be easily implemented in laboratories equipped with basic cell culture facilities and instrumentation for bioluminescence detection, that is, low-light detectors connected to inverted microscopes and luminometers, without the need for additional equipment.

Key words Bioluminescence, Spheroids, Imaging, Reporter gene, 3D cell models, Luciferase

1 Introduction

Bioluminescent (BL) cell-based assays based on two-dimensional (2D) cell cultures represent a well-established bioanalytical tool for interrogating natural biological phenomena, including gene expression, promoter regulation, and for monitoring molecular pathways, thus providing highly valuable information. Major applications of these assays are in drug discovery but also span from regenerative medicine to environmental monitoring and food control, as well as forensic applications [1–3]. Several cell-based assays have been developed based on bioluminescent reporters such as luciferases. Reporter gene technology relies on the splicing of transcriptional control elements to a variety of reporter proteins to monitor cellular events associated with gene expression and intracellular signaling cascades. The main advantages of these assays are wide dynamic range, high sensitivity, and suitability to high throughput-screening. Moreover, cell-based assays relying on luciferases as reporter protein are generally very simple to implement

Steven Ripp (ed.), *Bioluminescent Imaging: Methods and Protocols*, Methods in Molecular Biology, vol. 2081,
https://doi.org/10.1007/978-1-4939-9940-8_1,

and do not require expensive equipment or skilled personnel. More recently, thanks to the availability of luciferases emitting at different colors or requiring different substrates, multiplexed assays have also been developed to monitor two or more analytes or to include viability controls to correct the analytical signal according to cell viability [4, 5]. The recent technical advancements have dramatically reshaped the way cell-based assays are developed, expanding their applicability outside laboratories for on-site analyses. Several whole-cell biosensors have been reported in the past years, relying on the use of portable light detectors (such as charge-coupled devices and complementary metal oxide semiconductors) or simply integrated into 3D-printed smartphone-based devices [6–8]. These approaches are surely promising; however, 2D cell cultures do not always resemble the morphology and the functionality of the natural 3D environment. As a result, 2D cell culture tests can provide misleading data, reducing the reliability and predictivity of the results. Instead, 3D cell models are characterized by complex cell-to-cell communication and formation of extracellular matrix (ECM), thus restoring complex tissue architecture [9].

3D cell culture models represent an attractive approach to decode intra- and intercellular signaling, providing biologically relevant information and predictive data [10]. It is known that cells in 3D culture environments differ in gene, protein, and cell receptor expression from 2D-monolayer cell cultures and they provide an excellent model as surrogate in vivo systems. 3D structure spheroids exhibit enhanced cell viability, stable morphology and polarization, increasing proliferative activity and physiological metabolic function, which are markedly improved when compared to 2D cell monolayers [11, 12].

Besides the European Union, the USA and many other countries are strongly encouraging the implementation of in vitro assays and other approaches to replace animal testing, in accordance with the "three Rs principle" of Russell and Burch [13]. The possibility of exploiting bioluminescence detection for monitoring 3D cell models has been previously demonstrated by us and others [14–17].

We have previously reported a high-throughput BL 3D cell-based assay in micropatterned 96-well plate format using the transcriptional regulation of nuclear factor K beta response element in human embryonic kidney (HEK293) cells. Here we report a very straightforward method for bioluminescence imaging and bioluminescent reporter gene assays in 3D cell-culture models. Both the assays can be easily implemented in laboratories equipped with basic cell culture facilities and instrumentation for bioluminescence detection, that is, luminometer and low-light detectors connected to inverted microscopes, without the need for additional equipment.

2 Materials

2.1 3D Cell Culture and Transfection

1. Human embryonic kidney HEK293 cell line (ATCC® CRL-1573) or other cell lines that can be transfected readily.
2. Growth Medium: Dulbecco's modified Eagle's medium (DMEM) high glucose. Supplemented with 10% fetal bovine serum (FBS), 2 mM L-glutamine, 50 μg/mL penicillin and 50 μg/mL streptomycin (Carlo Erba reagents, Cornaredo, MI, Italy). Filter the growth medium using a filter cup 0.22 μm. Store at 4 °C (*see* **Note 1**).
3. Trypsin–EDTA solution 1× (Carlo Erba reagents, Cornaredo, MI, Italy). Store at 4 °C.
4. T25 flask culture and serological pipettes.
5. 96-well microspace round bottom cell culture plates (Elplasia™, Kuraray, Japan).
6. FuGENE HD transfection reagent (Promega, Madison, WI).
7. Plasmids: pGL4.32[PLG2] obtained by replacing Luc2P from vector pGL4.32[luc2P/NF-kB-RE/Hygro] (Promega, Madison, WI) with PLG2 luciferase gene [18]; plasmid for constitutive expression of the luciferase (e.g., pCMV_PLG2) to characterize luciferase emission in 3D cell cultures.

2.2 Reagents

1. 0.1 M phosphate buffered saline (PBS): Add 8 g of NaCl, 200 mg of KCl, 1.44 g of Na_2HPO_4, 240 mg of KH_2PO_4 into 1 L of double-distilled water. Adjust solution to desired pH (typically pH 7.4). Sterilize the solution in autoclave and prepare 50 mL aliquots. Store at room temperature.
2. 1.0 mM D-luciferin solution: Prepare Buffer n.1 (0.1 M citric acid solution) by dissolving 21.0 g of citric acid monohydrate, $C_6H_8O_7 \cdot H_2O$ into 1 L of distilled water; prepare Buffer n.2 (0.1 M trisodium citrate solution) by dissolving 29.41 g of trisodium citrate dihydrate, $C_6H_5O_7Na_3 \cdot 2H_2O$, into 1 L of distilled water. Mix 35 mL of Buffer n.1 and 65 mL Buffer n.2 for 30 min to obtain citric acid-sodium citrate buffer solution at pH 5.0. Dissolve 28.3 mg of D-luciferin into 100 mL of citric acid–sodium citrate buffer solution at pH 5.0. Prepare 8–10 mL aliquots and store at −20 °C (*see* **Note 2**).
3. Tumor Necrosis Factor Alpha (TNFα) stock solution: Centrifuge the vial before opening. Reconstitute the TNFα with 1 mL PBS sterile solution 0.1 M pH 7.4 containing 0.1% endotoxin-free recombinant human serum albumin in order to obtain a stock solution 10.0 μg/mL. Prepare 100 μL aliquots and store at −20 °C (*see* **Note 3**). Prepare on the day of the assay serial dilutions of TNFα (concentration range 0.1–50.0 ng/mL) using sterile deionized water.

2.3 Equipment and Software

1. For cell cultures: laminar flow cabinet and 5% CO_2 incubator maintained at 37 °C.
2. For bioluminescence imaging: inverted microscope (Olympus CK40) connected to an electron multiplying charge-coupled device (EMCCD) camera (ImagEM-X2, Hamamatsu). 10× (Olympus A10PL) and 4× (SPlan4SL, 0.16 N.A.) objectives are used for bright-field and BL imaging.
3. Light-tight box to protect the imaging system from light (*see* **Note 4**).
4. HCImage software (v 4.1.2.0, Hamamatsu Corporation) for image analysis.
5. For luminometric measurements: Varioskan Flash Luminometer (Thermo Fisher Scientific, Walthman, MA) and Thermo Scientific SkanIT Software for Varioskan Flash (version 2.4.3).
6. GraphPad Prism software v. 5.02 (GraphPad Software Inc.).
7. ImageJ version 1.51d software for spheroid analysis (https://imagej.nih.gov/ij/).

3 Methods

Carry out all procedures at room temperature (25 °C) under the laminar flow hood (unless indicated otherwise).

3.1 Spheroids Formation

1. Plate cells in 25 cm^2 flasks for cell culture in complete growth medium with serum. Culture cells overnight under standard conditions in preparation for spheroid production the following day.
2. Remove and discard the medium when the cells are approximately 80% confluent (80% of surface of 25 cm^2 flask covered by cell monolayer).
3. Gently rinse the cells with 5 mL of PBS 0.1 M pH 7.4 for removing all traces of serum that contains trypsin inhibitor.
4. Add 2.0 mL of prewarmed trypsin–EDTA solution 1× to the side wall of the flask. Observe the cells under inverted microscope. Detached cells appear rounded and float in the medium. If less than 80–90% of cells are detached incubate again the flask for an additional 2 min and observe the cells every 30 s. Then add 5 mL of complete fresh growth medium to stop the trypsin action (*see* **Note 5**).
5. Transfer the cell suspension to the tube and centrifuge the cells at 1600 × *g* for 8–10 min.

6. Remove the supernatant and add 5 mL of prewarmed complete growth medium, resuspend cells by gently pipetting.
7. Count the cells and plate them in a 96-well microspace round bottom Elplasia cell culture plate at a concentration of 2×10^4 cells per well in a total volume of 200 μL of medium (*see* **Note 6**).
8. Place the plate in incubator at 37 °C with 5% CO_2 for 48–72 h for spheroid formation (*see* **Note 7**).

3.2 Spheroid Analysis

1. Analyze bright-field images of HEK293 spheroids at desired time points (e.g., every 24 h) using ImageJ version 1.51d software to calculate the projected area (A) and perimeter (P) of each spheroid according to [19]. Calculate a sphericity factor (ϕ) as follows:

$$\phi = \frac{\pi \cdot \sqrt{\frac{4A}{\pi}}}{P}$$

2. Determine the projected area and perimeter for each spheroid using ImageJ and export data to MS EXCEL to calculate the sphericity factor of about 30 spheroids to obtain an average value ± standard deviation (*see* **Note 8**).
3. Calculate the average diameter of 30 spheroids, generally after 72 h of incubation (cell density of 4×10^4 cells) HEK293 spheroids with an average diameter of 210 ± 25 μm can be obtained (*see* **Note 9**).

3.3 Spheroid Transfection

Transfect 3-day-old HEK293 spheroids with 0.10 μg of plasmid per well of a 96-well Elplasia plate with Fugene®HD reagent with a ratio of 3:1 (*see* **Note 10**). Add a total volume of transfection mix (plasmid DNA and FuGENE® HD) of 2–10 μL per well. For transfecting 10 wells prepare the following mixture.

1. Mix by inverting briefly the Fugene HD reagent (*see* **Note 11**).
2. Add sterile deionized water to a sterile polystyrene Eppendorf to reach 62 μL final volume.
3. Add 1.1 μg plasmid in sterile H_2O and mix immediately by pipetting.
4. Add 3.3 μL Fugene HD reagent (ratio 3:1) and mix briefly by vortexing.
5. Incubate the Fugene HD reagent/DNA mixture for 5–10 min at room temperature.
6. Add 5 μL of this mixture per well and mix by gently moving the plate (*see* **Note 12**).
7. Place the plate in incubator at 37 °C, 5% CO_2 for 48 h.

3.4 Characterization of 3D HEK293 Expressing PLG2 Luciferase

Before performing the bioluminescent reporter gene assay and imaging sessions, characterize luciferase emission in spheroids in terms of emission wavelength and kinetics. These preliminary measurements are crucial to identify the optimal temporal window for BL measurements and for characterizing emission behavior in living cells. Transfect 1-day-old HEK293 spheroids with a plasmid for constitutive expression (plasmid pCMV_PLG2), 0.10 μg of per well with Fugene HD transfect reagent using the procedure previously described and incubate for 24 h at 37 °C, 5% CO_2. For luminometric measurements with Thermo Scientific SkanIT Software follow the procedure:

1. Create a new session file.
2. Select "*Plate layout*" and define wells to be measured.
3. Select "*Area*" and define the area for acquire BL signals in the 96-well plate.
4. Add "*Kinetic Loop*" for executing the steps multiple times before continuing with the protocol. Select number of readings of in the range 300–600.
5. In the "*Dispense*" step choose the dispenser, the volume of the substrate (100 μL) and medium dispensing speed. Use the "*Dispense at reading*" parameter for starting dispensing after 3 readings. While in the luminometric measurements.
6. Add "*Luminometric measurement*" for kinetic emission. Set the measurement time (range 200–1000 ms).
7. Add "Luminometric scanning" for emission spectrum. Select scanning wavelength start: 450 nm, end 700 nm; select step size 2 nm; dynamic range: autorange; measurement time: 200–1000 ms (*see* **Note 13**).
8. Add 100 μL of D-luciferin solution to each well with automatic injection and acquire kinetic measurements for 5–10 min integrating BL signals for 200 ms.
9. Export data and analyze results with GraphPad Prism software. Figure 1 shows a typical emission spectrum obtained with luciferase PLG2, characterized by lack of emission color change at low pH (<6.5) and high thermostability compared with wild-type *Photinus pyralis* luciferase (24 h versus 20 min at 37 °C) (*see* **Note 14**).

3.5 Bioluminescence Imaging of Spheroids

For performing bioluminescence imaging of HEK293 spheroids, transfect 3-day-old HEK293 spheroids with 0.10 μg of plasmid encoding a luciferase under regulation of constitutive promoter (e.g., pCMV_PLG2). Imaging session is performed after 48 h posttransfection with HEK293 cells, however, according to different cell types and transfection protocols the optimal window for BL measurements may vary from 24 to 72 h posttransfection.

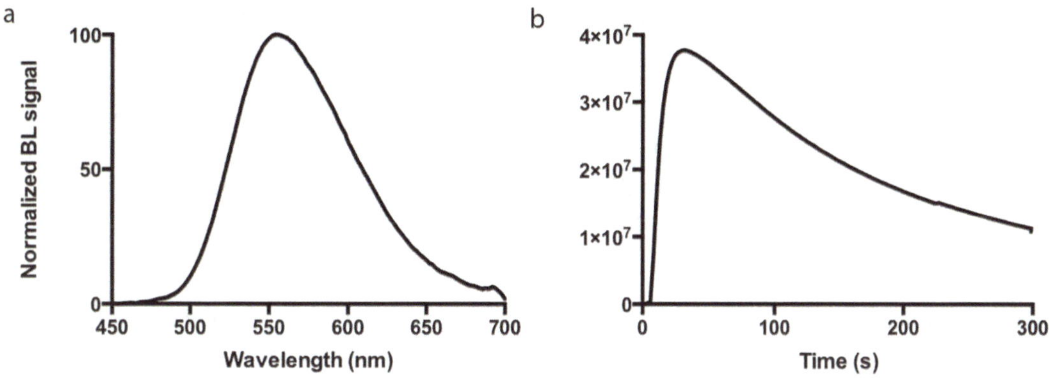

Fig. 1 (**a**) Normalized emission spectrum and (**b**) emission kinetic of PLG2 luciferase expressed in HEK293 spheroids obtained in nonlysing condition using D-luciferin buffer citrate substrate. BL signals are acquired with Varioskan Flash luminometer

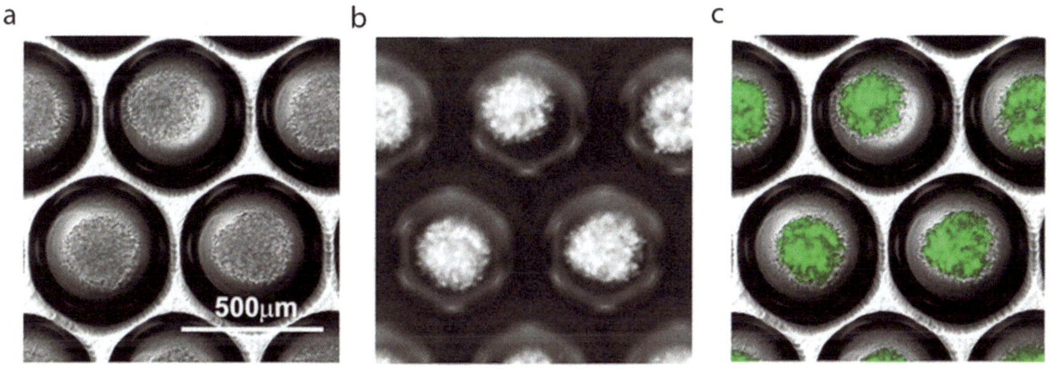

Fig. 2 Bright-field (a), bioluminescence (b), and pseudocolor overlay (c) images of HEK293 cells grown in 3D microplate format and expressing PLG2 luciferase under constitutive promoter regulation

1. Acquire bright-field images of the spheroids in the imaging field before acquiring bioluminescence imaging to localize the luminescence signals on the single spheroids (Fig. 2a).
2. Turn on the imaging system and the software on the computer (HCI images software). Wait for 30 min before image acquisition (*see* **Note 15**).
3. Place the plate on the microscope stage.
4. Set the EM gain at 4 and the exposure time at 30 ms.
5. Adjust the focus under the bright-field and capture the live image of spheroids using 10× and 4× objectives.
6. Add gently 100 μL of D-luciferin in citrate buffer pH 5.0, wait 30 s and acquire the BL imaging setting the EM gain level at 500 and the exposure time at 30 s (*see* **Note 16**).
7. Overlay images with HCI Images software (v 4.1.2.0) (*see* **Note 17**).

8. Use ImageJ version 1.51d software for calculating area (A) and the perimeter (P) of each spheroid in bright-field. For obtaining corrected BL emissions, quantify BL images designing manually a region of interest (ROI) around each spheroid and calculate BL intensities with respect to the projected area. Figure 2b shows an example of BL imaging with bright-field, bioluminescence, and pseudocolor overlay images of HEK293 cells grown in 3D and expressing PLG2 luciferase.

3.6 Bioluminescent Imaging and Luminometric Assay for Inflammatory Pathway Activation

As proof of concept of 3D bioluminescence imaging of HEK293 spheroids, an inflammatory reporter gene assay is described. Transfect 3-day-old HEK293 spheroids with 0.10 μg of plasmid pNF-kB_PLG2 encoding PLG2 luciferase under the control of the transcriptional regulation of the nuclear factor k beta (NF-kB) response element. After 48 h posttransfection, HEK293 spheroids are incubated with TNFα solutions (concentration range 0.1–50 ng/mL) for 4 h at 37 °C, 5% CO_2 (Fig. 3). Cell viability is monitored by transfecting separate wells with plasmid pCMV_PLG2 or by cotransfecting with a luciferase emitting at a different wavelength if filters for spectral resolution are available. The detailed protocol is here described:

1. Change the medium adding fresh growth medium until the well is completely filled.
2. Remove gently 240 μL of medium in order to obtain 100 μL of volume per well.
3. Add 50 μL of TNFα solutions in culture medium in order to obtain final concentration range from 0.1 to 20 ng/mL. Add 50 μL of culture medium in the control wells. Perform all experiments at least in triplicate and repeat them at least three times.
4. Incubate the plate for 4 h at 37 °C, 5% CO_2.

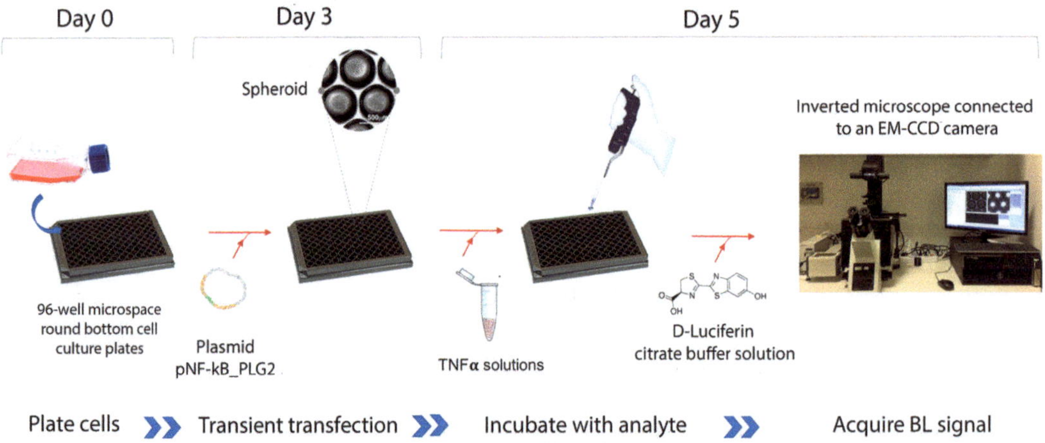

Fig. 3 Schematic representation of bioluminescence-based imaging assay for inflammatory pathway activation obtained with HEK293 spheroids transfected with pNF-kB_PLG2 vector

For the luminometric assay:

1. Acquire the BL emission with Varioskan Flash by creating a new session with Thermo Scientific SkanIT Software as previously described in the Subheading 3.4 setting the parameters for luminometric measurements with automatic injection of 100 μL of D-luciferin.
2. Analyze data with GraphPad Prism software and calculate the fold induction as the ratio between the average relative light units of induced cells and average relative light units of control cells and analyze data with GraphPad Prism software.

For the imaging assay:

1. Perform imaging of 3D cell cultures with the ImagEMX2 EMCCD camera using 4× (SPlan4SL) objectives with an integration time of 30 s, at a gain level set to 500, after the addition of 100 μL of 1 mM D-luciferin substrate per well (*see* **Note 18**).
2. Manually design the region of interest around each spheroid and calculate BL intensity with respect to the projected area (corrected BL emission).
3. Calculate the mean of corrected BL emissions obtained from 20 spheroids to construct the dose–response curve generated using GraphPad Prism software (Fig. 4).

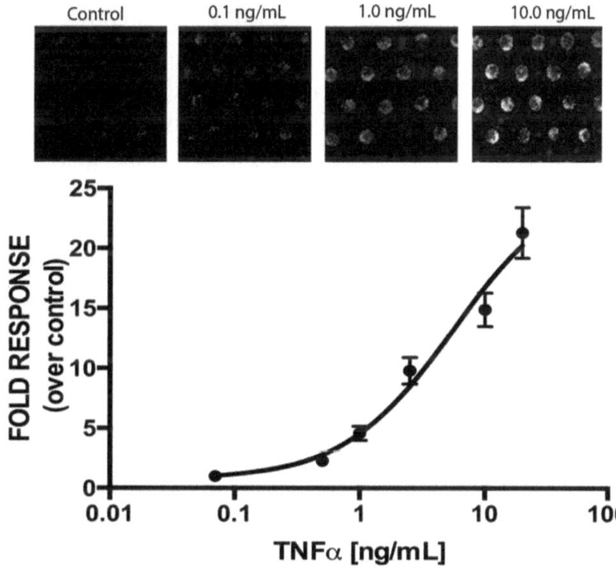

Fig. 4 Dose–response curve and BL images obtained with HEK293 spheroids transfected with 0.10 μg of reporter plasmid pNF-kB_PLG2 and incubated with different concentrations of TNFα for 4 h at 37 °C. BL signals are obtained after the addition of nonlysing D-luciferin buffer citrate substrate

4 Notes

1. Warm cell culture media to 37 °C prior to use.
2. D-luciferin must be protected from light. Cover the aliquots with aluminum foil before storage at −20 °C.
3. Lyophilized TNF remains active for 1 year at −20 °C. Upon reconstitution, it can be stored at 2–8 °C for short term only, or at −20 °C to −80 °C in aliquots for long term storage. Avoid repeated freeze–thaw cycles.
4. A customized light-tight box can be fabricated instead of commercial ones.
5. To avoid clumping do not agitate the cells by hitting or shaking the flask while waiting for the cells to detach.
6. Before seeding add 100 μL of complete culture medium to each well, then add 200 μL of cell suspension.
7. Since spheroid formation may vary depending on cell type, check the cells every 24 h.
8. Average values of sphericity factor obtained with Elplasia plates are generally close to 1, meaning that the aggregates are of nearly spherical shape.
9. Select the incubation time according to desired spheroid size which must be in accordance to the goal of the experiment. In order to obtain spheroid without necrosis in the core and to maintain the functionality of the cells within the aggregate an average diameter of 150–200 μm is suggested. Besides, oxygen availability at the core of spheroids represents an issue for luciferase-based assays.
10. Transfection efficiency may vary depending on cell type and transfection reagent. Change transfection reagent–DNA ratio to improve transfection efficiency. Transfection efficiency can be estimated by cotransfecting with a plasmid expressing a green fluorescent protein and measuring the percentage of fluorescent cells.
11. Before use, allow the vial of the transfection reagent to reach room temperature.
12. Before transfection, add fresh growth medium to completely fill the wells, then remove 245 μL of medium in order to obtain a final volume of 100 μL after transfection.
13. The "*Measurement time*" defines the light collection time used in the measurement. For luminometric intensities a default measurement time of 1000 ms is suggested; however, when luciferases with high specific activity are used lower measurement times can be selected (e.g., 200 ms).

14. When bioluminescence measurements are performed within living cells it is of crucial importance to record emission spectra in order to assess eventual changes in luciferase emissions due to low pH or high temperature. In particular, when two luciferases emitting at different colors are measured in the same well by spectral resolution pH- and temperature-insensitive luciferases must be selected to avoid emission overlapping.
15. Before imaging, let the EM-CCD camera reach a temperature of −65 °C. A stable cooling temperature is essential to obtain superior performance and for stable gain settings.
16. Addition of reagents is a critical step and liquid reagents should be gently added to leave spheroids undisturbed.
17. Acquire the bright-field images before and after the addition of D-luciferin substrate to verify the correct focus.
18. The choice of objectives is critical for detecting bioluminescent images since the light collection power is proportional to the square of numerical aperture and indirectly proportional to the magnification value square.

References

1. Cali JJ, Niles A, Valley MP, O'Brien MA, Riss TL, Shultz J (2008) Bioluminescent assays for ADMET. Expert Opin Drug Metab Toxicol 4:103–120
2. Michelini E, Cevenini L, Mezzanotte L, Coppa A, Roda A (2010) Cell-based assays: fuelling drug discovery. Anal Bioanal Chem 398:227–238
3. Cevenini L, Calabretta MM, Calabria D, Roda A, Michelini E (2016) Luciferase genes as reporter reactions: how to use them in molecular biology? Adv Biochem Eng Biotechnol 154:3–17
4. Michelini E, Cevenini L, Mezzanotte L, Ablamsky D, Southworth T, Branchini BR, Roda A (2008) Combining intracellular and secreted bioluminescent reporter proteins for multicolor cell-based assays. Photochem Photobiol Sci 7(2):212–217
5. Tarnow P, Bross S, Wollenberg L, Nakajima Y, Ohmiya Y, Tralau T, Luch A (2017) A novel dual-color luciferase reporter assay for simultaneous detection of estrogen and aryl hydrocarbon receptor activation. Chem Res Toxicol 30 (7):1436–1447
6. Cevenini L, Calabretta MM, Tarantino G, Michelini E, Roda A (2016) Smartphone-interfaced 3D printed toxicity biosensor integrating bioluminescent "sentinel cells". Sensors Actuators B Chem 225:249–257
7. Li H, Lopes N, Moser S, Sayler G, Ripp S (2012) Silicon photomultiplier (SPM) detection of low-level bioluminescence for the development of deployable whole-cell biosensors: possibilities and limitations. Biosens Bioelectron 33(1):299–303
8. Cevenini L, Calabretta MM, Lopreside A, Tarantino G, Tassoni A, Ferri M, Roda A, Michelini E (2016) Exploiting NanoLuc luciferase for smartphone-based bioluminescence cell biosensor for (anti)-inflammatory activity and toxicity. Anal Bioanal Chem 408 (30):8859–8868
9. Bahrvand H, Hashemi SM, Kazemi Ashtini S, Farrokhi A (2006) Differentiation of human embryonic stem cells into hepatocytes in 2D and 3D culture systems in vitro. Int J Dev Biol 50:645–652
10. Antoni D, Burckel H, Josset E, Noel G (2015) Three-dimensional cell culture: a breakthrough in vivo. Int J Mol Sci 16(3):5517–5527
11. Benya PD, Shaffer JD (1982) Dedifferentiated chondrocytes reexpress the differentiated collagen phenotype when cultured in agarose gels. Cell 30:215–224
12. Nelson CM, Bissell MJ (2005) Modeling dynamic reciprocity: engineering three-dimensional culture models of breast architecture, function, and neoplastic transformation. Semin Cancer Biol 15:342–352

13. Michelini E, Cevenini L, Calabretta MM, Calabria D, Roda A (2014) Exploiting in vitro and in vivo bioluminescence for the implementation of the three Rs principle (replacement, reduction, and refinement) in drug discovery. Anal Bioanal Chem 406 (23):5531–5539
14. Yasunaga M, Fujita Y, Saito R, Oshimura M, Nakajima Y (2017) Continuous long-term cytotoxicity monitoring in 3D spheroids of beetle luciferase-expressing hepatocytes by nondestructive bioluminescence measurement. BMC Biotechnol 17(1):54. https://doi.org/10.1186/s12896-017-0374-1
15. Suzuki T, Kanamori T, Inouye S (2017) Quantitative visualization of synchronized insulin secretion from 3D-cultured cells. Biochem Biophys Res Commun 486(4):886–892
16. Cevenini L, Calabretta MM, Lopreside A, Branchini BR, Southworth TL, Michelini E, Roda A (2017) Bioluminescence imaging of spheroids for high-throughput longitudinal studies on 3D cell culture models. Photochem Photobiol 93(2):531–535
17. Michelini E, Calabretta MM, Cevenini L, Lopreside A, Southworth T, Fontaine DM, Simoni P, Branchini BR, Roda A (2019) Smartphone-based multicolor bioluminescent 3D spheroid biosensors for monitoring inflammatory activity. Biosens Bioelectron 123:269–277. https://doi.org/10.1016/j.bios.2018.09.012
18. Branchini BR, Southworth TL, Fontaine DM, Kohrt D, Talukder M, Michelini E, Cevenini L, Roda A, Grossel MJ (2015) An enhanced chimeric firefly luciferase-inspired enzyme for ATP detection and bioluminescence reporter and imaging applications. Anal Biochem 484:148–153
19. Kelm JM, Timmins NE, Brown CJ, Fussenegger M, Nielsen LK (2003) Method for generation of homogeneous multicellular tumor spheroids applicable to a wide variety of cell types. Biotechnol Bioeng 83:173–180

Chapter 2

Bioluminescence Resonance Energy Transfer (BRET) Coupled Near-Infrared Imaging of Apoptotic Cells

Setsuko Tsuboi and Takashi Jin

Abstract

Detection of apoptotic cells is crucial for understanding the mechanism of diseases and for therapy development. So far, visible-emitting fluorescent probes such as FITC-labeled Annexin V has been widely used for the detection of apoptotic cells. However, such probes cannot be applied to noninvasive imaging in the near-infrared (NIR) region. Compared with visible light, NIR light is highly permeable in turbid biological samples and tissues. In addition, NIR optical imaging has several advantages such as lower autofluorescence and scattering from biological samples, leading to clearer images with high signal to background ratios. Here, we describe the synthesis and application of bioluminescence resonance energy transfer (BRET)-coupled quantum dots (QDs) for the NIR optical imaging of apoptotic cells.

Key words Bioluminescence resonance energy transfer (BRET), Renilla luciferase, Near-infrared imaging (NIR imaging), Quantum dots (QDs), Annexin V, Apoptotic cell

1 Introduction

Programmed cell death (apoptosis) plays an important role not only in the homeostasis of normal tissues but also in the pathological conditions of diseased tissues [1, 2]. The detection of apoptotic cells is crucial for understanding the mechanism of those diseases and for therapy development. Deregulation in apoptosis induces numerous diseases such as cancer and cardiovascular and neurodegenerative diseases [2, 3]. During the early process of apoptosis [4], morphological changes of the cell membrane induce the asymmetrical distribution of phosphatidylserine (PS) on the outer monolayer of the membrane [5, 6]. Annexin V is an endogenous protein having binding ability to PS in the presence of Ca^{2+} ions, and this protein is widely used as a probe for the detection of apoptotic cells [7]. So far, for the detection of apoptotic cells, Annexin V has been modified with fluorescence, luminescence, magnetic resonance contrast, and radioisotopes [7]. Among these probes, fluorescent-dye labeled probes such as FITC- and Cy3/Cy5-Annexin V [8, 9]

Steven Ripp (ed.), *Bioluminescent Imaging: Methods and Protocols*, Methods in Molecular Biology, vol. 2081,
https://doi.org/10.1007/978-1-4939-9940-8_2,

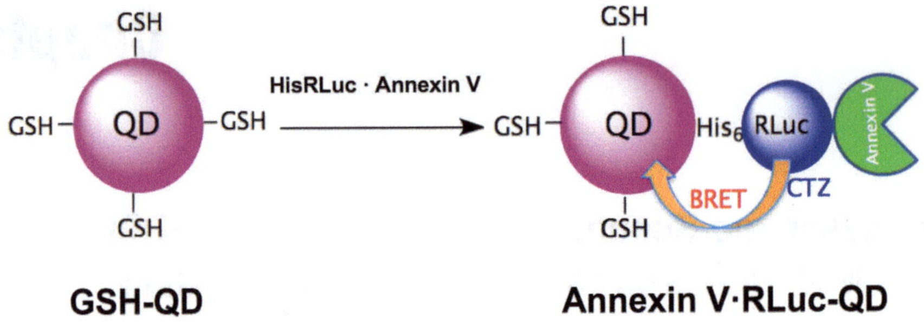

Fig. 1 Schematic representation for the preparation of a recombinant protein, HisRLuc-Annexin V functionalized QDs (Annexin V·RLuc-QDs), where HisRLuc·Annexin V protein directly binds to the surface of GSH-QDs. BRET occurs from CTZ to QD in an Annexin V·RLuc-QD

are the most popular for the optical detection of apoptotic cells by using fluorescence microscopy or fluorescence activated cell sorting. However, most fluorescence-labeled Annexin V-probes emit at the visible region less than 700 nm. Such visible-emitting probes encounter high background signals arising from autofluorescence of intrinsic fluorescent species in biological samples.

We have developed a NIR optical imaging technique for the detection of apoptotic cells using bioluminescence resonance energy transfer (BRET) [10–12] coupled annexin V functionalized NIR-QDs [13, 14]. Compared with fluorescence imaging, BRET-based luminescence imaging does not require external excitations, leading to low autofluorescence from biological samples. In this chapter, we describe the synthesis and application of recombinant Annexin V protein functionalized QDs for BRET-coupled NIR optical detection of apoptotic cells (Fig. 1). This probe is designed to emit NIR luminescence due to BRET from *Renilla luciferase* (RLuc) [15]-coelenterazine (CTZ) to glutathione (GSH) coated NIR emitting CdSeTe/CdS QDs (GSH-QDs) [16, 17].

2 Materials

2.1 Synthesis of NIR-Emitting QDs

1. Selenium (powder, 99.999%).
2. Tellurium (shot, 1–2 mm, 99.99%).
3. Cadmium 2,4-pentanedionate (ODPA).
4. Tri-octylphosphine oxide (TOPO).
5. Trioctylphosphine (TOP).
6. Tri-butylphosphine (TBP).
7. Hexadecylamine (HDA, 90%).
8. Sulfur (S, crystalline, 99.999%).
9. Inert gas (nitrogen).
10. Fluorescence spectrometer.

2.2 Surface Modification of QDs with Glutathione (GSH)

1. Glutathione (GSH, reduced form).
2. Potassium *t*-butoxide.
3. Gel filtration column.
4. Aqueous solution of 10 mM Na_2CO_3.

2.3 Protein Synthesis (HisRLuc-Annexin V)

1. Plasmids containing the sequence of Renilla Mullerei Luciferase (RLuc) and Annexin V.
2. Bacterial vector for high-level expression of proteins with 6×His tag, such as pRSET plasmid (ThermoFisher).
3. PCR primers, high-fidelity DNA polymerase, PCR buffer, dNTPs, nuclease-free water and thermal cycler.
4. TAE buffer: 40 mM Tris, 20 mM acetic acid, 1 mM EDTA.
5. 1% Agarose gel: 1% agarose (w/v), TAE buffer.
6. Ethidium Bromide solution (0.5 μg/mL).
7. Gel extraction kit.
8. Cloning enzymes such as ligase and restriction enzymes.
9. Competent cells for protein expression, SOC medium.
10. LB plates with antibiotic appropriate for the plasmid.
11. Plasmid extraction kit.
12. Sequencing system.
13. LB medium, antibiotics, L-Rhamnose, isopropyl β-D-1-thiogalactopyranoside (IPTG).
14. Binding Buffer: 50 mM Tris, 500 mM NaCl, 20 mM imidazole.
15. Elution Buffer: 50 mM Tris, 500 mM NaCl, 500 mM imidazole.
16. Protease inhibitor.
17. Ni Sepharose 6 Fast Flow (GE Healthcare).
18. Gel filtration column.
19. Phosphate-Buffered Saline (PBS).
20. Ultrasonic homogenizer, Ultracentrifuge, Rotator, Spectrophotometer.
21. SDS-PAGE apparatus. 5–20% polyacrylamide gel, Tris-glycine-SDS buffer, Coomassie Brilliant Blue.

2.4 Bioluminescence Spectral Measurements

1. Aqueous solution of HisRLuc-Annexin V (1 mg/mL in PBS).
2. Ethanol solution (1 mg/mL) of coelenterazine (CTZ).
3. Phosphate-Buffered Saline (PBS).
4. Luminescence spectrometer.

2.5 Protein Conjugation to GSH-QDs

1. Aqueous solution of HisRLuc-Annexin V (1 mg/mL in PBS).
2. Aqueous solution of GSH-QDs (1 μM in 10 mM Na_2CO_3, O.D. = ca. 3 at 488 nm).
3. PBS.
4. Gel filtration column.

2.6 BRET Spectra of Annexin V·RLuc-QDs

1. Aqueous solution of Annexin V·RLuc-QDs (1 μM in PBS).
2. PBS.
3. Ethanol solution (1 mg/mL) of coelenterazine (CTZ).
4. Luminescence spectrometer.

2.7 Cell Preparation and Apoptosis Induction

1. Desired cell line for biologic question of interest.
2. General apparatus and reagents used for cell culture. Medium, culture dish, incubator, microscope, and so on.
3. Actinomycin D (ActD).

2.8 Flow Cytometric Analysis and Cell Imaging

1. 1× Annexin V binding buffer (2.5 mM $CaCl_2$ in medium).
2. FITC-Annexin V.
3. Propidium iodide (PI) solution.
4. DMSO solution of ActD.
5. Flow cytometer.

2.9 BRET-Coupled Luminescence Imaging of Apoptotic Cells

1. Aqueous solution of Annexin V·RLuc-QDs (1 μM in PBS).
2. Ethanol solution (1 mg/mL) of CTZ.
3. Luminescence imaging system.

3 Methods

NIR-emitting CdSeTe/CdS (core/shell) QDs are prepared by a high temperature decomposition method [16, 17]. A recombinant Annexin V protein (HisRLuc·Annexin V) was prepared by the transformation of pRSET-RLuc-Annexin V plasmid into *E. coli.* HisRLuc·Annexin V protein has six histidine tags which directly bind to the surface of GSH capped CdSeTe/CdS QDs (Fig. 1). This is due to the high affinity of histidine to Cd^{2+} ions at the surface of a Cd-S shell layer [18–20]. The GSH capped CdSeTe/CdS QDs (GSH-QDs) show a broad absorption spectrum from NIR to visible region (400–800 nm) to be able act as an acceptor for the CTZ emission (around 485 nm).

3.1 QD Synthesis (CdSeTe/CdS)

Carry out all procedures for QD synthesis in a fume hood, unless otherwise specified.

3.1.1 Se-Te Stock Solution

1. Dissolve Se (24 mg, 0.3 mmol) and Te (13 mg, 0.1 mmol) in TBP (1 mL) at room temperature (*see* **Note 1**).
2. Preserve the Se-Te stock solution under a nitrogen atmosphere.

3.1.2 CdSeTe Core

1. Load a mixture of cadmium 2,4-pentanedionate (150 mg, 0.48 mmol), ODPA (300 mg, 0.90 mmol), TOPO (1 g), HDA (3 g), and TOP (0.5 mL) into a 25 mL three-necked flask and heated to 330 °C under a nitrogen atmosphere.
2. At this temperature, 1 mL of a Se-Te stock solution was quickly injected by using a syringe, which causes an immediate change in the solution color from colorless to brown.
3. By monitoring the fluorescence spectra, check the formation of CdSeTe QDs (ca. 790 nm emission) (*see* **Note 2**). After the CdSeTe QDs were formed, cool the solution to 60 °C, and add chloroform (10 mL) to the solution.
4. Precipitate CdSeTe QDs by the addition of methanol.
5. Separate the QDs by centrifugation (*see* **Note 3**).
6. Dissolve the QDs in 2 mL of TBP.

3.1.3 Cd-S Stock Solution

1. Dissolve sulfur (40 mg, 1.25 mmol) in TBP (10 mL) at 100 °C. After sulfur is completely dissolved, cool the solution to room temperature.
2. Add cadmium 2,4-pentanedionate (388 mg, 1.25 mmol) to the sulfur solution, and warm the solution at 100 °C to dissolve cadmium 2,4-pentanedionate.
3. Preserve the Cd-S stock solution under a nitrogen atmosphere.

3.1.4 CdS Overcoating

1. Load a TBP solution of CdSeTe QDs and HDA (3 g) into a 25 mL three-necked flask and heat to 250 °C.
2. Add a Cd-S stock solution (0.25 mL) to the solution of CdSeTe/CdS QDs.
3. Check the formation of QDs (ca. 830 nm emission) by monitoring the fluorescence spectra. After the QDs are formed, cool the solution to 50 °C.
4. Add chloroform (10 mL) to the solution.
5. Precipitate CdSeTe/CdS QDs by addition of methanol.
6. Separate CdSeTe/CdS QDs by centrifugation (*see* **Note 3**).
7. Dissolve the CdSeTe/CdS QDs into 20 mL of tetrahydrofuran.

3.2 Surface Modification of QDs with Glutathione (GSH)

1. Add 1 mL of an aqueous solution (50 mg/mL of GSH) to a tetrahydrofuran solution (2 mL) of CdSeTe/CdS QDs (1 μM, O.D. = ca. 3 at 488 nm) at room temperature under sonication.
2. Separate QD precipitates by centrifugation.
3. Add 2 mL of an aqueous solution of potassium *t*-butoxide (20 mg/1 mL) to the QD precipitates.
4. Sonicate the solution for 5 min.
5. Filter the solution through a 0.45 μm membrane filter to remove excess GSH and potassium *t*-butoxide.
6. Dialysis the solution with an aqueous solution of Na_2CO_3 (10 mM) using a 50,000 MW membrane (*see* **Note 4**).
7. Measure the fluorescence spectra and transmission electron microscope image of purified GSH-QDs (Fig. 2).

3.3 Protein Synthesis (HisRLuc·Annexin V)

1. Insert the DNA fragment of RLuc and Annexin V amplified by PCR into a bacterial vector.
2. Transform the constructed HisRLuc·Annexin V plasmid into competent cells for protein expression.
3. Grow precultures overnight in LB medium containing antibiotic appropriate for the plasmid at 37 °C.
4. Dilute overnight precultures 1:100 into LB containing antibiotic appropriate for the plasmid and grow cultures at 37 °C until their density (O.D.600) reach to 0.5–0.6.
5. When the cultures reach an O.D.600 of 0.5–0.6, shift temperature to 15–25 °C and induce protein expression by adding

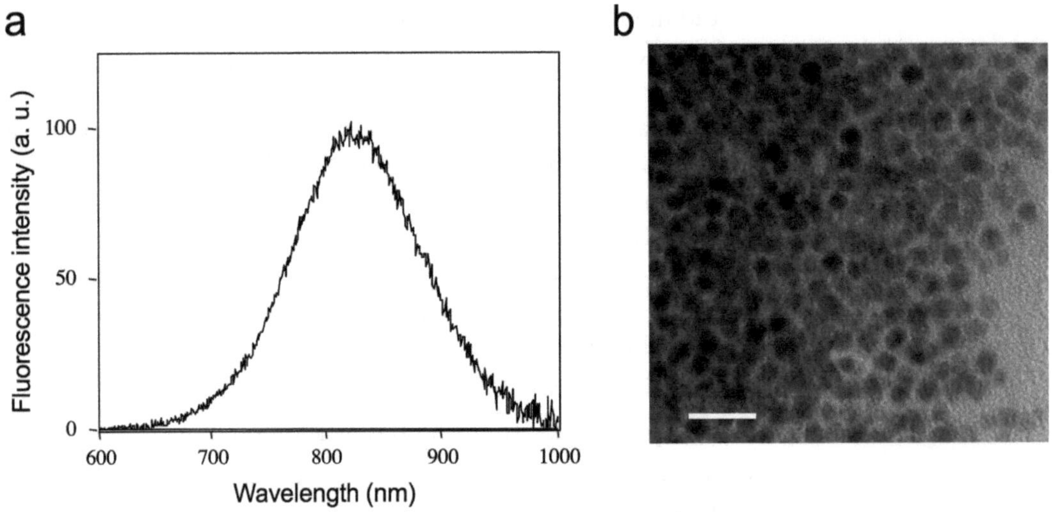

Fig. 2 (**a**) Fluorescence spectra of GSH-QDs in water. (**b**) Transmission electron spectra of GSH-QDs. Scale bar: 20 nm

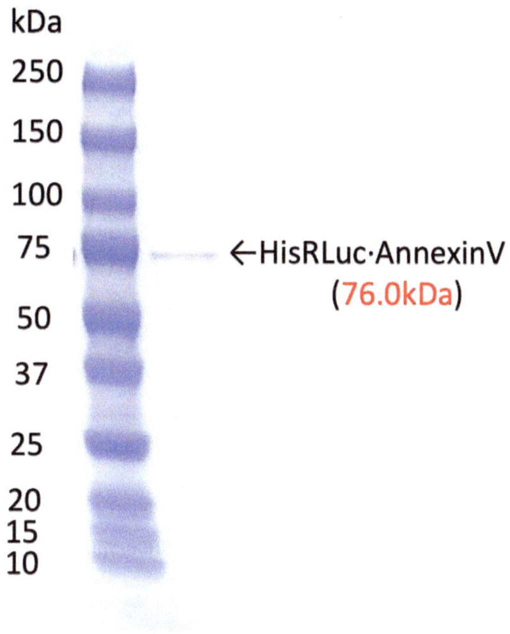

Fig. 3 SDS-PAGE of HisRLuc-Annexin V

isopropyl β-D-1-thiogalactopyranoside (0.2 mM) and L-Rhamnose (0.1%).

6. Grow the cultures overnight at 15–25 °C with shaking at 190 rpm.
7. Collect the cells by centrifugation at 5000 × *g* for 10 min.
8. Resuspend the cell pellet in binding buffer with protease inhibitor added just prior to use.
9. Sonicate the cell suspension using a 10 s bursts at middle-intensity with a cooling period of 10 s between each burst on ice.
10. Centrifuge the lysate at 20,000 × *g* for 20 min to eliminate cell debris.
11. Purify according to the instructions of Ni Sepharose 6 Fast Flow.
12. Exchange the buffer of eluted fractions with PBS using a gel filtration column.
13. Confirm the purified HisRLuc·Annexin V protein by SDS-PAGE (Fig. 3). A size marker is used to compare the molecular weight of HisRLuc·Annexin V protein. The expected molecular size for HisRLuc·Annexin V protein was 76 kDa.

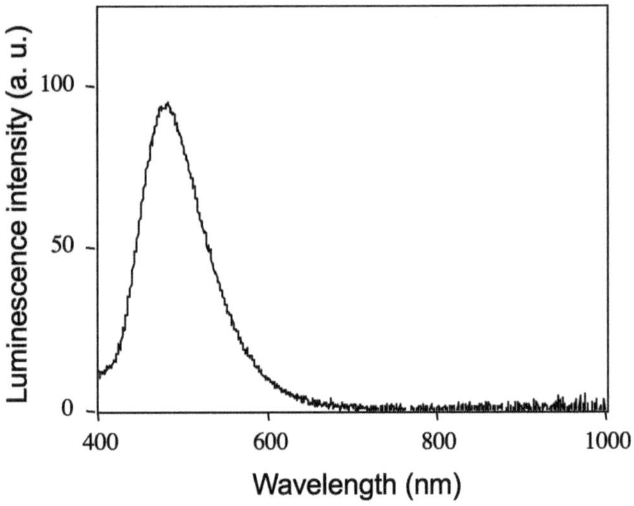

Fig. 4 Bioluminescence spectra for HisRLuc-Annexin V

3.4 Bioluminescence Spectra for HisRLuc·Annexin V

1. Add 20 μL of an aqueous solution of HisRLuc·Annexin V (1 mg/mL) to 2 mL of PBS.
2. Add 10 μL of an ethanol solution (1 mg/mL) of CTZ.
3. Measure bioluminescence spectra using a luminescence spectrometer (Hamamatsu Photonics) with 1 min exposure time (Fig. 4, *see* **Note 5**).

3.5 Protein Conjugation to GSH-QDs

1. Add 300 μL of HisRLuc·Annexin V (1 mg/mL in PBS) to 0.4 mL of GSH-QDs (1 μM in 10 mM Na_2CO_3 solution).
2. Exchange the solution buffer with PBS by using a gel filtration column (PD-10 column, GE Healthcare).
3. Check the purity of HisRLuc·Annexin V conjugated QDs by agarose gel electrophoresis (Fig. 5a).

3.6 BRET Spectra of Annexin V·RLuc-QDs

1. Add 20 μL of an aqueous solution of Annexin V·RLuc-QDs (1 μM) to 2 mL of PBS.
2. Add 10 μL of an ethanol solution (1 mg/mL) of CTZ.
3. Measure BRET spectra using a luminescence spectrometer (Hamamatsu Photonics) with 1 min exposure time (Fig. 5b) (*see* **Note 5**).

3.7 Cell Preparation and Apoptosis Induction

1. Culture a cell line in a suitable medium for 24 h.
2. Optimal induction conditions are found by adding aqueous solutions of ActD (0–10 μg/mL) in the target cell line for 24 h and use it as a positive control (*see* **Note 6**).

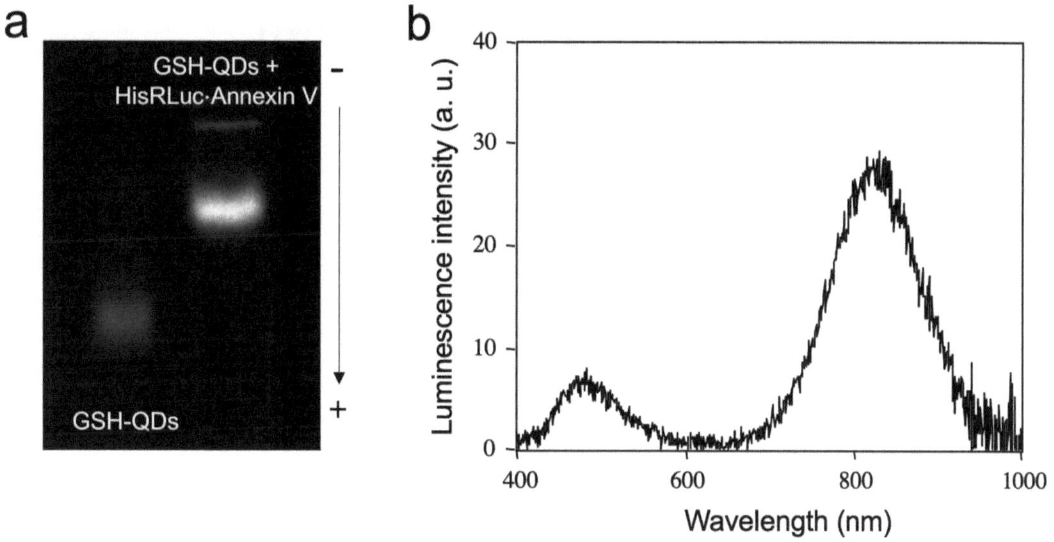

Fig. 5 (**a**) Agarose gel electrophoresis of GSH-QDs and the mixture of GSH-QDs/ HisRLuc-Annexin V (1:10). (**b**) BRET spectra of Annexin V·RLuc-QDs in the presence CTZ

3.8 Flow Cytometric Analysis and Cell Imaging

1. Collect the culture medium in a 15 mL conical tube.
2. Wash the cells with PBS, collect this wash solution and combine with the collected culture medium from **step 1**.
3. Collect cells by trypsinization and mix with the cell suspension collected in **steps 1** and **2**.
4. Pellet these cells by centrifugation (300 × *g*, 3 min), discard the supernatant, resuspend the cell pellet in 1× Annexin V binding buffer at a concentration of ~10^6 cells/mL.
5. Transfer 100 μL of the solution to a 1.5 mL sample tube.
6. Add 5 μL of an aqueous solution of FITC-Annexin V and 5 μL an aqueous solution of PI. Incubate for 15 min at room temperature in the dark.
7. Add 400 μL of 1× Annexin V binding buffer to each tube and pass through a 40-μm cell strainer mesh.
8. Analyze by flow cytometry as soon as possible (Fig. 6) (*see* **Note 7**).
9. Transfer a portion of the cell suspension used for flow cytometric analysis to a 35 mm glass bottom dish (D 11134 H, Matsunami) and observe the fluorescence of FITC-Annexin V an FITC filter (Fig. 7).

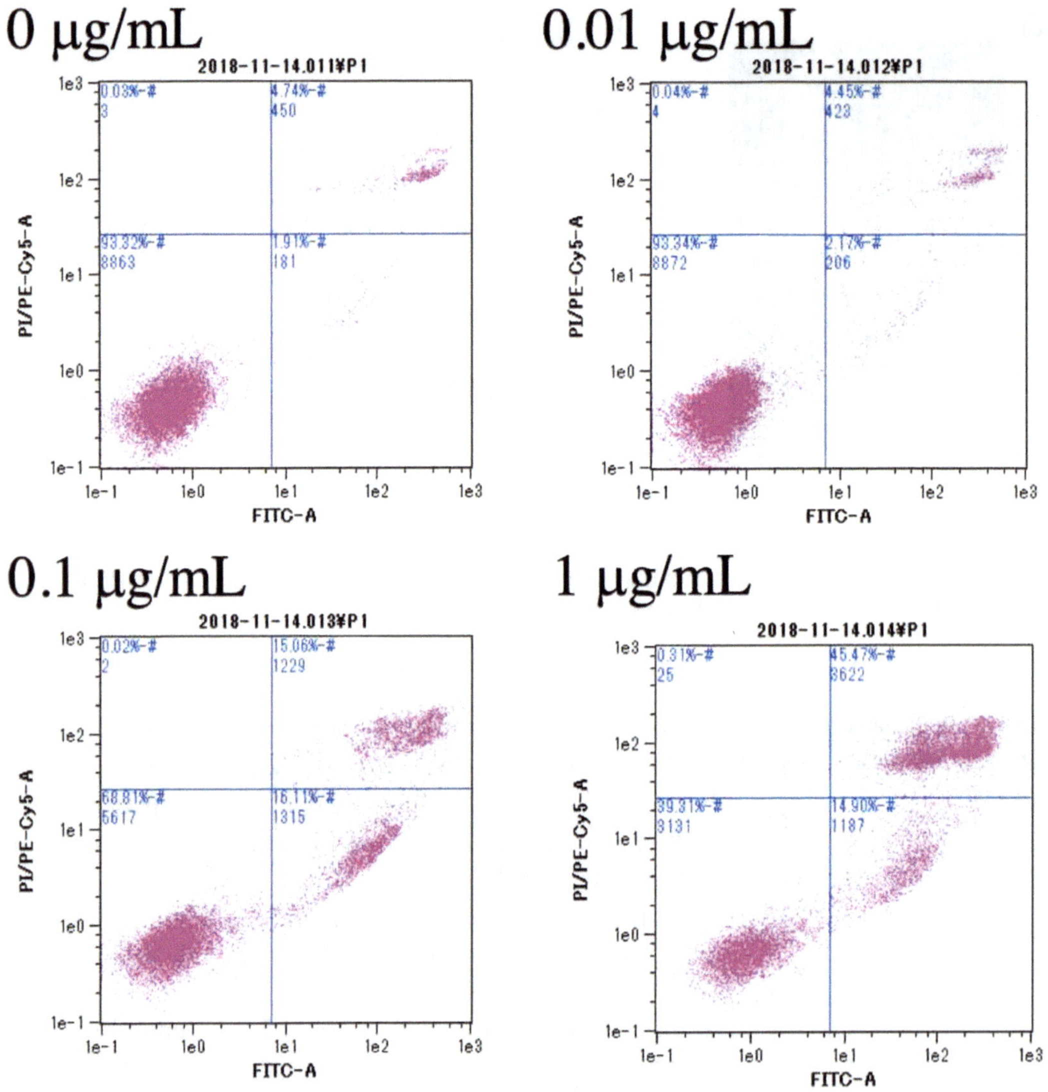

Fig. 6 Flow cytometric analysis for HeLa cells treated without and with ActD. Images were taken after incubation with ActD (0, 0.01, 0.1, and 1 μg/mL) for 24 h. Green and red fluorescence shows the emission form FITC-Annexin V and PI, respectively

3.9 BRET-Coupled Luminescence Imaging of Apoptotic Cells

1. Add Annexin V·RLuc-QDs (final concentration of 30 nM) to the cell suspension collected similarly to flow cytometry (*see* **Note 8**).
2. Incubate for 15 min at room temperature in the dark.
3. Wash the cell suspension with 1× Annexin V binding buffer three times and resuspend the cells with 50 μL of PBS.
4. Immediately before the detection of their NIR luminescence, add CTZ (final concentration of 50 μM) to the cell suspension.

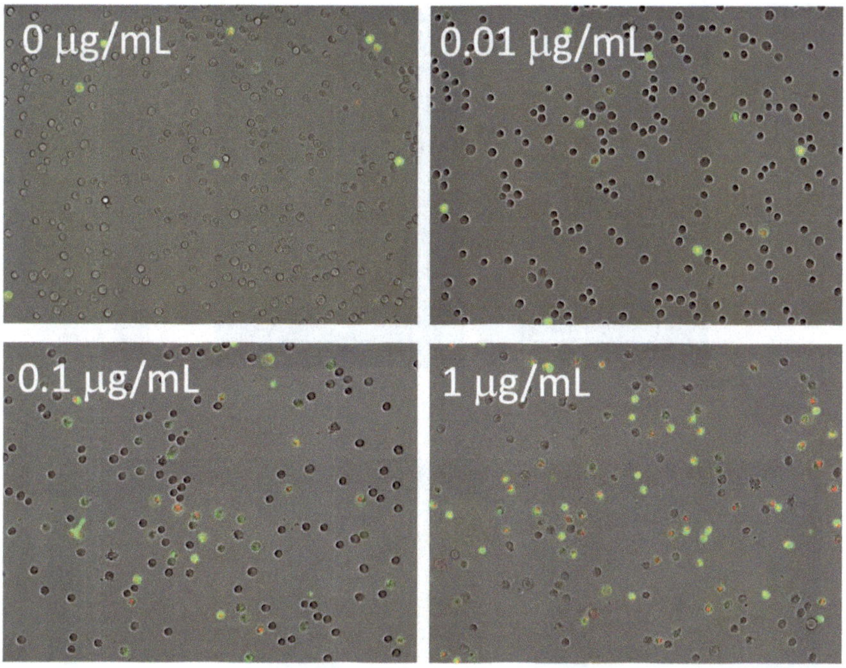

Fig. 7 Cellular imaging of HeLa cells treated without and with ActD. Images were taken after incubation with ActD for 24 h. Green and red fluorescence shows the emission from FITC-Annexin V and PI, respectively

5. Take BRET-coupled NIR images (at 830 ± 20 nm) by using a luminescence imaging system (Bruker, MS FX PRO) with 10 min exposure time (Fig. 8).

4 Notes

1. Use a freshly prepared solution of TBP. When the Se and Te are not completely dissolved in TBS, sonicate the TBP solution.
2. For this purpose, we use a fluorescence spectrometer nearby a fume food and monitor the fluorescence spectra of QDs.
3. For the separation of QDs, we employ test (glass) tubes for centrifugation. Control the speed of centrifugation so as not to damage the test (glass) tubes.
4. For the exchange of the aqueous solution of QDs with Na_2CO_3 (10 mM), a gel filtration column (PD-10 column, GE Healthcare) can also be used.
5. To obtain bioluminescence spectra for HisRLuc·Annexin V, the spectra should be measured immediately after the addition of CTZ to the aqueous solution of HisRLuc·Annexin V.
6. In the case of HeLa cells, Actinomycin D (1 µg/mL) is added to induce apoptosis in about 60% of the cells by induction for 24 h.

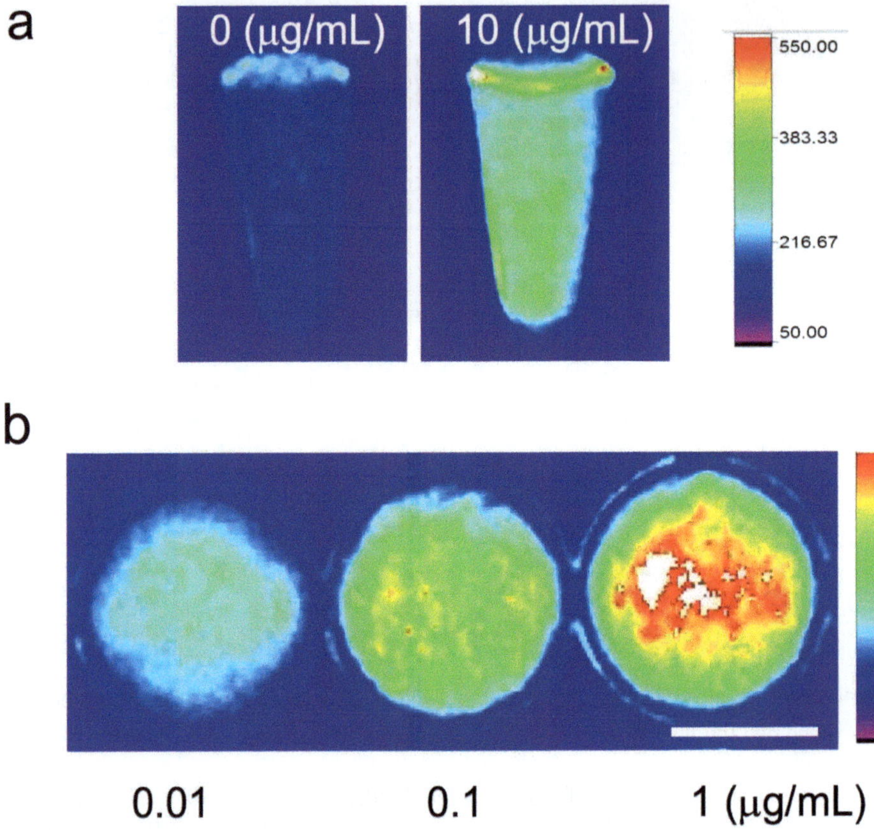

Fig. 8 BRET coupled NIR imaging of HeLa cells (~10^6 cells) after incubation with Annexin V·RLuc-QDs. (**a**) HeLa cells treated without and with ActD (10 μg/mL). (**b**) HeLa cells treated with 0.01, 0.1, and 1 μg/mL of ActD. Scale bar: 10 mm

7. Flow cytometric analysis is performed using the MACSQuant Analyzer (Miltenyi Biotec Inc.). PI is collected through a FL4 filter (ex: 488 nm, em: 655–730 nm band-pass), FITC-Annexin V fluorescence is collected through a FL2 (FITC) filter (ex: 488 nm, em: 525 ± 50 nm).
8. Just before use, mix 1 μM of GSH-QDs and 1 mg/mL of HisRLuc·Annexin V at a volume ratio of 2:1.

References

1. Taylor RC, Cullen SP, Martin SJ (2008) Apoptosis: controlled demolition at the cellular level. Nat Rev Mol Cell Biol 9:231–241
2. Fuchs Y, Steller H (2015) Live to die another way: modes of programmed cell death and the signals emanating from dying cells. Nat Rev Mol Cell Biol 16:329–344
3. Elmore S (2007) Apoptosis: a review of programmed cell death. Toxicol Pathol 35:495–516
4. Fuchs Y, Steller H (2011) Programmed cell death in animal development and disease. Cell 147:742–758

5. van Genderen HO, Kenis H, Hofstra L, Narula J, Reutelingsperger CP (2008) Extracellular annexin A5: functions of phosphatidylserine-binding and two-dimensional crystallization. Biochim Biophys Acta 1783:953–963
6. Lizarbe MA, Barrasa JI, Olmo N, Gavilances F, Turnay J (2013) Annexin-phospholipid interactions. Functional implications. Int J Mol Sci 14:2652–2683
7. Nazari M, Minai-Tehrai A, Emamzadeh R (2014) Comparison of different probes based on labeled annexin V for detection of apoptosis. RSC Adv 4:45128–45135
8. Koopman G, Reutelingsperger C, Kuijten G, Keehnen R, Pals S, Van Oers M (1994) Annexin V for flow cytometric detection of phosphatidylserine expression on B cells undergoing apoptosis. Blood 84:1415–1420
9. Petrovsky A, Schellenberger E, Josephson L, Weissleder R, Bogdanov A Jr (2003) Near-infrared fluorescent imaging of tumor apoptosis. Cancer Res 63:1936–1942
10. Hasegawa M, Tsukasaki Y, Ohyanagi T, Jin T (2013) Bioluminescence resonance energy transfer coupled near-infrared quantum dots using GST-tagged luciferase for *in vivo* imaging. Chem Commun 49:228–230
11. Samanta A, Walper SA, Susumu K, Dwyer CL, Medinz IL (2015) An enzymatically-sensitized sequential and concentric energy transfer relay self-assembled around semiconductor quantum dots. Nanoscale 7:7603–7614
12. Yu X, Wen K, Wang Z, Zhang X, Li C, Zhnag S, Shen J (2016) General bioluminescence resonance energy transfer homogeneous immunoassay for small molecules based on quantum dots. Anal Chem 88:3512–3520
13. Kamkaew A, Sun H, England CG, Cheng L, Liu Z, Cai W (2016) Quantum dot–NanoLuc bioluminescence resonance energy transfer enables tumor imaging and lymph node mapping *in vivo*. Chem Commun 52:6997–7000
14. Weissleder R (2001) A clearer vision for in vivo imaging. Nat Biotechnol 19:316–317
15. Srikantha T, Klapach A, Lorenz WW, Tsai LK, Laughlin LA, Gorman JA, Soll DR (1996) The sea pansy *Renilla reniformis* luciferase serves as a sensitive bioluminescent reporter for differential gene expression in *Candida albicans*. J Bacteriol 178:121–129
16. Jin T, Yoshioka Y, Fujii F, Komai Y, Seki J, Seiyama A (2008) Gd^{3+}-functionalized near-infrared quantum dots for *in vivo* dual modal (fluorescence/magnetic resonance) imaging. Chem Commun 44:5764–5766
17. Jin T, Fujii F, Komai Y, Seki J, Seiyama A, Yoshioka Y (2008) Preparation and characterization of highly fluorescent, glutathione-coated near infrared quantum dots for in vivo fluorescence imaging. Int J Mol Sci 9:2044–2061
18. Medintz IL, Uyeda HT, Goldman ER, Mattoussi H (2005) Quantum dot bioconjugates for imaging, labelling and sensing. Nat Mater 4:435–446
19. Lao UL, Mulchandani A, Chen W (2006) Simple conjugation and purification of quantum dot-antibody complexes using a thermally responsive elastin-protein L scaffold as immunofluorescent agents. J Am Chem Soc 128:14756–14757
20. Alam R, Karam LM, Doane TL, Zylstra J, Fontaine DM, Branchini BR, Maye MM (2014) Near infrared bioluminescence resonance energy transfer from firefly luciferase-quantum dot bionanoconjugates. Nanotechnology 25:495606. (7pp)

Chapter 3

High-Throughput Analysis of Endocrine-Disrupting Compounds Using BLYES and BLYAS Bioluminescent Yeast Bioassays

Tingting Xu, Anna Young, Jasleen Narula, Gary Sayler, and Steven Ripp

Abstract

Bioluminescent yeast assays BLYES and BLYAS are whole-cell bioassays that utilize genetically modified *Saccharomyces cerevisiae* bioreporters to detect estrogenic and androgenic activities, respectively. The bioreporter strains chromosomally express human estrogen receptor alpha (BLYES) or androgen receptor (BLYAS) and contain a reporter plasmid expressing the complete bacterial luciferase gene cassette (*luxCDABE*) under the control of an estrogen- or androgen-responsive promoter. Exposure to endocrine-disrupting compounds activates the receptor which subsequently turns on the expression of the reporter genes, resulting in dose-dependent bioluminescence (i.e., light) emission. These yeast whole-cell bioassays provide rapid, cost-effective, and high-throughput detection of endocrine-disrupting activities in environmental samples. This protocol will provide a detailed description of the standard assay procedures as well as a framework for data analysis.

Key words Endocrine-disrupting compounds, Emerging contaminants, Bioluminescence, Bacterial luciferase, Bioreporter, High-throughput screening

1 Introduction

Endocrine-disrupting compounds (EDCs) are a group of emerging contaminants that are of substantial animal and human health concern due to their potential to interfere with natural endocrine signaling activities. The vertebrate endocrine system utilizes small signaling molecules (hormones) to transcriptionally regulate the expression of a sophisticated gene network throughout different stages of an animal's life cycle including development, growth and reproduction. EDCs, however, can mimic or repress native hormone functions, resulting in an imbalance in the delicate regulation of the endocrine network in humans and wildlife. Mounting scientific evidence has demonstrated that exposure to EDCs can lead to reproductive, developmental, and metabolic disorders inclusive of breast, ovarian, testicular, and prostate cancers, early onset of

Steven Ripp (ed.), *Bioluminescent Imaging: Methods and Protocols*, Methods in Molecular Biology, vol. 2081,
https://doi.org/10.1007/978-1-4939-9940-8_3,

puberty, reduced sperm counts, and various other adverse health effects [1, 2]. Unfortunately, many compounds that make up everyday consumer products such as plastics, household and personal care products, pharmaceuticals, and pesticides have been found to be EDCs, and their wide use and discharge into the environment through anthropogenic activities and industrial waste has made monitoring of EDCs in the aqueous ecosystem a critical checkpoint to safeguard the health of humans and wildlife.

Analytical methods such as gas chromatography/mass spectrometry are the traditional gold standards for EDC detection. However, despite their supreme sensitivity, analytical methods are expensive, time-consuming, require specialized instruments and personnel, and most importantly, lack the capability to predict EDC bioavailability. As a complement to analytical methods, a variety of whole-cell bioassays have been developed to provide a biological evaluation of the cumulative effects of EDCs in environmental samples without identifying individual compounds. Using the lower eukaryotic organism *Saccharomyces cerevisiae* as a platform for EDC screening, two bioluminescent yeast assays BLYES [3] and BLYAS [4] have been developed to detect estrogenic and androgenic activity, respectively. These genetically engineered *S. cerevisiae* strains chromosomally express a hormone-specific nuclear receptor (i.e., human estrogen receptor alpha in BLYES and human androgen receptor in BLYAS), which upon activation by EDC exposure will turn on the episomal expression of a reporter operon under the control of a hormone-specific promoter. In particular, BLYES and BLYAS express the complete *Photorhabdus luminescens* luciferase gene cassette (*luxCDABE*) that allows for autonomous bioluminescent production in a dose-dependent manner without the need of exogenous substrate addition or photonic excitation. A constitutively bioluminescent strain BLYR [5] has also been developed and is often used in parallel with BLYES and/or BLYES strains to assess the toxicity of environmental samples.

These yeast bioassays can be conveniently performed in test tubes for low-throughput applications or in microtiter plates (i.e., 96-well plates) for simultaneous analysis of multiple samples in a single assay. They provide rapid (usually within a few hours), cost-effective, and high-throughput detection of EDCs and have been successfully applied worldwide for chemical screening and evaluation [6–9] and EDC monitoring in surface and drinking water [10, 11] and wastewater [12, 13]. The bioluminescent signal is easily detectable by conventional luminometers for low-throughput applications, or by microplate readers with luminescence capability for high-throughput assays. Here we present the detailed protocol of using the BLYES, BLYAS, and BLYR bioluminescent yeast bioreporters in a 96-well plate format to analyze EDC contaminants in environmental samples.

2 Materials

2.1 Yeast Strains

Bioluminescent yeast bioreporter strains used in this protocol include *Saccharomyces cerevisiae* BLYES for estrogenicity detection, BLYAS for androgenicity detection, and BLYR for toxicity evaluation [3, 4, 6].

2.2 Glassware

Glassware must be used for medium preparation and storage and yeast growth. Wrap glassware openings with aluminum foil and sterilize by baking at 500 °C for 4 h in a muffle furnace (*see* **Note 1**).

2.3 Chemical Standards

Serial dilutions of 17β-estradiol (E2) and 5α-dihydrotestosterone (DHT) are used as positive controls and calculation standards for BLYES estrogenicity and BLYAS androgenicity detection, respectively. Methanol (HPLC grade) is used as solvent and negative control in BLYES and BLYAS assays. Chemical standards are prepared as below.

1. Dissolve 13.67 mg E2 in 5 mL of methanol in a borosilicate glass vial to make a 1×10^{-2} M E2 stock.
2. Serially dilute the 1×10^{-2} M E2 stock by tenfold in separate glass EPA vials with methanol to final concentrations of 1×10^{-3} M, 1×10^{-4} M, 1×10^{-5} M, 1×10^{-6} M, 1×10^{-7} M, 1×10^{-8} M, 1×10^{-9} M, 1×10^{-10} M, and 1×10^{-11} M.
3. Dilute the 1×10^{-6} M E2 stock by twofold with methanol to 5×10^{-7} M.
4. Perform 5 tenfold serial dilutions of the 5×10^{-7} M E2 solution in separate glass EPA vials with methanol to final concentrations of 5×10^{-8} M, 5×10^{-9} M, 5×10^{-10} M, 5×10^{-11} M, and 5×10^{-12} M.
5. Dilute the 5×10^{-7} M E2 solution by twofold with methanol in a glass vial to 2.5×10^{-7} M.
6. Perform 5 tenfold serial dilutions of the 2.5×10^{-7} M E2 solution in separate glass EPA vials with methanol to final concentrations of 2.5×10^{-8} M, 2.5×10^{-9} M, 2.5×10^{-10} M, 2.5×10^{-11} M, and 2.5×10^{-12} M.
7. The final full series of E2 standards will include 18 dilutions, ranging from 1×10^{-6} M to 2.5×10^{-12} M.
8. To prepare DHT standards, first dissolve 14.52 mg DHT into 5 mL methanol to generate a 1×10^{-2} M stock.
9. From the 1×10^{-2} M DHT stock, follow similar dilution patterns in Subheading 2.3 **steps 2–6** to generate a full series of DHT standards consisting of 18 dilutions ranging from 1×10^{-5} M to 2.5×10^{-11} M (*see* **Note 2**).
10. Store E2 and DHT standards at 4 °C.

Table 1
Yeast minimal medium (YMM) without leucine, uracil, or tryptophan for the BLYES, BLYAS, and BLYR assay

Component	Concentration
Ammonium sulfate ($(NH_4)_2SO_4$)	1.7 g/L
Copper(II) sulfate ($CuSO_4$)	12 mg/L
Iron(II) sulfate ($FeSO_4$)	684 μg/L
Monopotassium phosphate (KH_2PO_4)	11.6 g/L
Potassium hydroxide (KOH)	3.6 g/L
Magnesium sulfate ($MgSO_4$)	171 μg/L
D-(+)-glucose	20 g/L
Biotin	20 μg/L
Pantothenic acid	400 μg/L
Inositol	1 mg/L
Pyridoxine	400 μg/L
Thiamine	400 μg/L
Adenine	42.7 mg/L
Arginine–HCl	17.1 mg/L
Aspartic acid	100 mg/L
Glutamic acid	85.5 mg/L
Histidine	42.73 mg/L
Isoleucine	25.64 mg/L
Lysine–HCl	25.64 mg/L
Methionine	17.1 mg/L
Phenylalanine	21.4 mg/L
Serine	320.4 mg/L
Threonine	192 mg/L
Tyrosine	25.7 mg/L

See Subheading 2.4 for preparation steps

2.4 Medium Preparation

All yeast strains should be grown in yeast minimal medium (YMM) without leucine, uracil, or tryptophan for optimal detection sensitivity (Table 1).

1. Prepare a 0.8 g/L $FeSO_4$ and a 20 mM $CuSO_4$ solution with HPLC-grade water. Sterilize each by filtration through a 0.2 μm filter membrane into separate sterile, baked glass bottles. Store at room temperature.

2. Prepare 10 g/L L-tyrosine, 4 g/L L-aspartic acid, and 24 g/L L-threonine solutions with HPLC water. Filter-sterilize each into separate baked glass bottles and store at room temperature.
3. To make the amino acid mixture, mix 1.3 g L-histidine, 1.3 g adenine, 0.52 g L-arginine–HCl, 0.52 g L-methionine, 0.78 g L-isoleucine, 0.78 g L-lysine–HCl, 0.65 g L-phenylalanine, 2.6 g L-glutamic acid, 3.9 g L-valine, and 9.75 g L-serine with 1 L HPLC water. Sterilize by filtering through a 0.2 μm filter into a baked glass bottle and store at room temperature.
4. In a baked glass bottle, mix 13.61 g KH_2PO_4, 1.98 g $(NH_4)_2SO_4$, 4.2 g KOH pellets, and 0.2 g $MgSO_4$ with 953.5 mL HPLC water. Sterilize by autoclaving at 121 °C for 20 min. Add 1 mL $FeSO_4$ solution (0.8 g/L) (Subheading 2.4, **step 1**), 3 mL L-tyrosine solution (10 g/L) (Subheading 2.4, **step 2**), and 38.5 mL amino acid mixture (Subheading 2.4, **step 3**) to bring the total volume to 1 L. This completes the preparation of yeast base medium.
5. Prepare a 200 g/L D-(+)-glucose solution in HPLC water in a baked glass bottle. Sterilize by autoclave at 121 °C for 10 min. Store at room temperature.
6. The vitamin solution contains 40 mg/L thiamine, 40 mg/L pyridoxine, 40 mg/L pantothenic acid, 200 mg/L inositol, and 2 mg/L biotin in HPLC water. Sterilize by filtering through a 0.2 μm filter into a baked glass bottle and store at 4 °C.
7. To make the final YMM yeast assay medium, combine 2.5 mL $CuSO_4$ solution (prepared in Subheading 2.4, **step 1**), 25 mL L-aspartic acid and 8 mL L-threonine (Subheading 2.4, **step 2**), 854.5 mL yeast base medium (Subheading 2.4, **step 4**), 100 mL glucose solution (Subheading 2.4, **step 5**), and 10 mL vitamin solution (Subheading 2.4, **step 6**) to a final volume of 1 L and filter through a 0.2 μm filter membrane into a baked glass bottle. Store at room temperature for routine use or at 4 °C for long term (>1 month).

3 Methods

3.1 Preparation of Environmental Samples

1. Extract 1 L aqueous sample into 1 mL HPLC-grade methanol using solid phase extraction U.S. EPA method 1694 to generate a 1000× concentrate.
2. From the 1000× concentrate, follow similar dilution patterns in Subheading 2.3 (**steps 2–6**) to generate a full scale of sample

extract dilutions consisting of 12 concentrations ranging from 1000× to 0.25×.

3. Store sample methanol dilutions at 4 °C.

3.2 Yeast Bioassay

1. Thaw a 1 mL-vial frozen stock (−80 °C) of strain BLYES, BLYAS, and BLYR at room temperature (*see* **Note 3**).
2. Immediately following thawing, inoculate the cells into 30 mL YMM without leucine, uracil, or tryptophan in a baked 250 mL glass flask.
3. Incubate at 28 °C and 200 rpm shaking to an approximate optical density at 600 nm (OD_{600}) of 0.5–0.8 (*see* **Note 4**).
4. Pipet 20 μL of each of the 18 E2 or DHT standards into individual wells of an opaque 96-well microtiter plate for BLYES estrogenicity or BLYAS androgenicity detection, respectively (Fig. 1).
5. Add 20 μL methanol into three separate wells as negative controls for each assay (*see* **Note 5**) (Fig. 1).

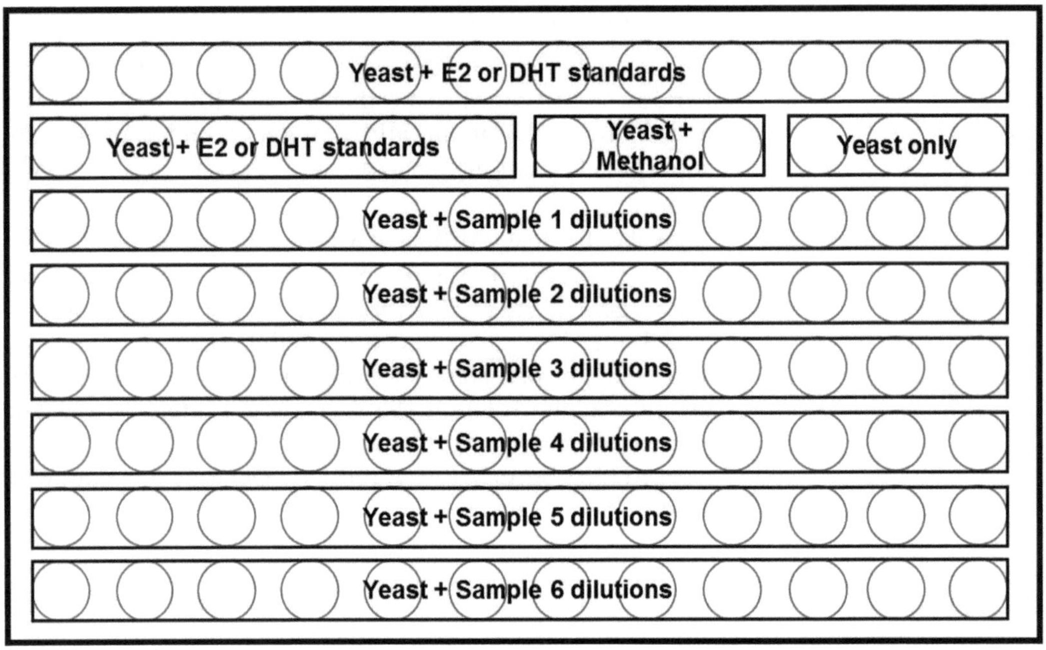

Fig. 1 An example 96-well plate layout for BLYES and BLYAS assays. Preload 20 μL of each standard dilution, 20 μL methanol, and 100 μL of each sample dilution to corresponding wells shown on the plate. Include three empty wells as blanks. Allow methanol to fully evaporate prior to adding 200 μL of yeast culture to each well of the plate (including the empty wells). BLYR assay is similarly plated, except that no chemical standards are needed

6. Add 100 μL of each of the 12 sample methanol extract dilutions into individual wells of an opaque 96-well microtiter plate (Fig. 1).
7. Allow the methanol to evaporate until no remaining visible wet spot. This usually takes 30 min to 1 h.
8. Add 200 μL BLYES culture into each of the wells preloaded with E2 standards and samples of interest for estrogenicity assay. Avoid bubbles.
9. Pipet 200 μL BLYAS culture into each of the wells preloaded with DHT standards and samples of interest for androgenicity assay.
10. Add 200 μL BLYR culture into each of the sample wells for toxicity monitoring.
11. Add 200 μL BLYES, BLYAS, or BLYR culture to triplicate wells preloaded with only methanol as negative controls for each assay.
12. Final assay concentrations of E2, DHT, and sample extracts after mixing with yeast culture range from 1×10^{-7} M to 2.5×10^{-13} M, 1×10^{-6} M to 2.5×10^{-12} M, and 500× to 0.125×, respectively.
13. Add 200 μL BLYES or BLYAS cells to three additional empty wells as blanks to monitor potential contamination of the methanol solvent (*see* **Note 6**).
14. Seal the plate with a breathable-sealing membrane.
15. Transfer the plate into a microplate reader and measure bioluminescence using a 1 s/well integration time every hour for 8 h at 28 °C (*see* **Note 7**).
16. Use bioluminescence readings collected 4–8 h post treatment for analysis (*see* **Note 8**).

3.3 Calculation of IC_{20} for Toxicity Evaluation

1. Inhibitory concentration (IC_{20}) is used to evaluate sample toxicity and is defined as the sample concentration capable of reducing the bioluminescence by 20% compared to methanol only controls in BLYR. Using an example dataset (Table 2) for demonstration, this section provides a step-by-step tutorial of toxicity calculation.
2. Calculate the average bioluminescence from BLYR controls exposed to only methanol.
3. Calculate the relative bioluminescence from sample-treated BLYR cells by dividing the measured bioluminescence by the calculated average methanol control readings.
4. Plot the relative bioluminescence against logarithmically transformed sample concentration factors (log(CF)) (Fig. 2).

Table 2
An example data set of BLYES estrogenicity, BLYAS androgenicity, and BLYR toxicity test of environmental samples

Sample dilutions	Concentration factor (×)	BLYR (RLU)	BLYES (RLU)	BLYAS (RLU)
	500	216	342	4133
	250	249	209	2639
	125	255	191	1514
	50	287	170	707
	25	295	283	857
	12.5	286	265	539
	5	328	222	522
	2.5	352	198	549
	1.25	348	187	559
	0.5	337	166	521
	0.25	341	168	523
	0.125	331	163	501

Methanol control	BLYR (RLU)	BLYES (RLU)	BLYAS (RLU)
	347	175	526
	340	190	498
	341	183	496

Chemical standards	[E2] (M)	BLYES (RLU)	[DHT] (M)	BLYAS (RLU)
	1×10^{-7}	3934	1×10^{-6}	8263
	5×10^{-8}	4211	5×10^{-7}	8314
	2.5×10^{-8}	3947	2.5×10^{-7}	10,713
	1×10^{-8}	4779	1×10^{-7}	9799
	5×10^{-9}	4016	5×10^{-8}	8248
	2.5×10^{-9}	3174	2.5×10^{-8}	8104
	1×10^{-9}	2979	1×10^{-8}	7921
	5×10^{-10}	1608	5×10^{-9}	5172
	2.5×10^{-10}	842	2.5×10^{-9}	3090
	1×10^{-10}	414	1×10^{-9}	922
	5×10^{-11}	220	5×10^{-10}	516
	2.5×10^{-11}	212	2.5×10^{-10}	515
	1×10^{-11}	183	1×10^{-10}	565
	5×10^{-12}	188	5×10^{-11}	478
	2.5×10^{-12}	179	2.5×10^{-11}	477
	1×10^{-12}	192	1×10^{-11}	494
	5×10^{-13}	176	5×10^{-12}	482
	2.5×10^{-13}	190	2.5×10^{-12}	481

RLU, relative light unit

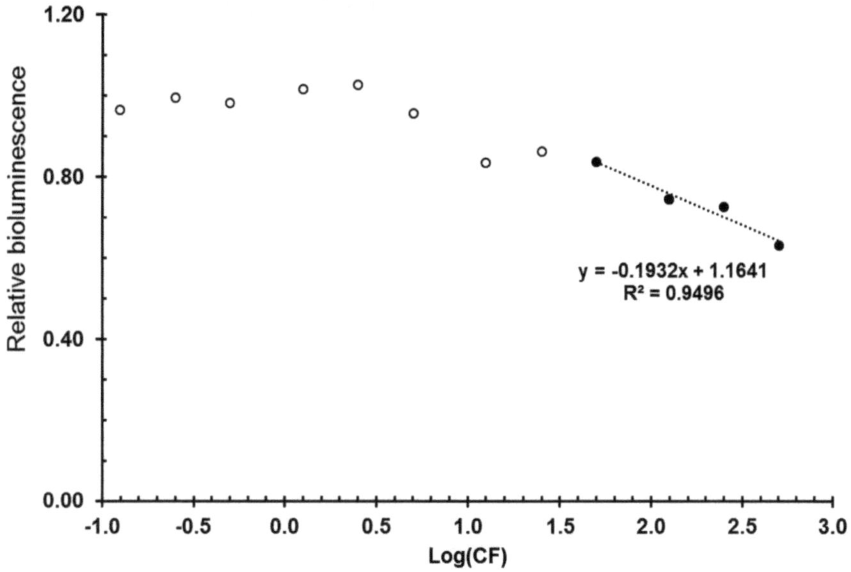

Fig. 2 BLYR toxicity assay data analysis. Relative bioluminescence is plotted against logarithmically transformed sample concentration factors (log(CF)) for the full scale of sample dilutions (open circle). Data points from the linear portion of the curve (closed circle) are used for linear regression prediction

5. Choose data points from the linear portion of the curve and fit the data with a linear regression model:

$$\text{Relative bioluminescence} = \text{intercept} + \text{slope} \times \log(\text{CF})$$

In this example, data points for the four highest concentrations are used for analysis. A linear relationship between relative bioluminescence and log(CF) is predicted as:

$$\text{Relative bioluminescence} = 1.1641 - 0.1932 \times \log(\text{CF})$$

Calculate IC_{20} with a relative bioluminescence value of 0.8 and other parameters in the predicted linear model. The IC_{20} of the sample in this example is calculated to be 76.7×, meaning that the sample needs to be concentrated 76.7 times to cause a 20% reduction of bioluminescence in BLYR.

3.4 Calculation of Equivalent Estrogenicity and Androgenicity

Sample estrogenicity and androgenicity is expressed as equivalent E2 and DHT concentration, respectively, which are calculated based on the bioreporter response to chemical standards. Using an example dataset (Table 2) for demonstration, this section provides a step-by-step tutorial of equivalent estrogenicity and androgenicity calculation.

1. Plot the bioluminescent readings against the E2 and DHT standards assay concentrations (Fig. 3).

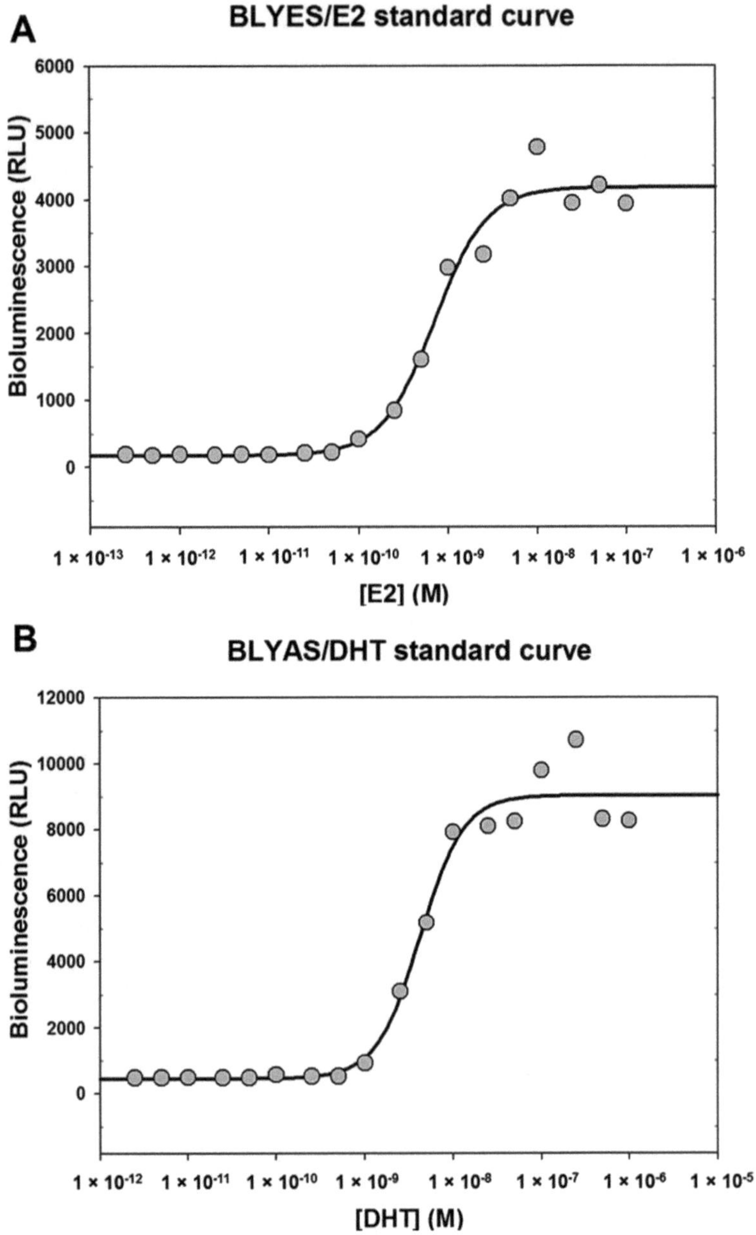

Fig. 3 BLYES and BLYAS chemical standard curve. Sigmoidal dose–response curves are predicted using a four-parameter logistic regression model for E2 (**a**) and DHT (**b**) in BLYES and BLYAS assay, respectively. RLU, relative light unit

2. Fit the data points using a four-parameter logistic regression model to generate a chemical standard curve:

$$\text{Bioluminescence} = \min + (\max - \min)/\left[1 + (\text{concentration}/\text{EC}_{50})^{(-\text{hillslope})}\right]$$

In this model, bioluminescence and concentration are variables, whereas "min," "max," "EC_{50}," and "hillslope" are parameters that can be solved. In this example, the EC_{50} value of E2 in the BLYES assay and DHT in the BLYAS assay is determined to be 7.0×10^{-10} M and 4.2×10^{-9} M, respectively, and the fitted standard curve can be expressed as:

$$\text{BLYES bioluminescence} = 176.5 + (4182.7 - 176.5)/\left[1 + \left([\text{E2}]/(7.0 \times 10^{-10})\right)^{(-1.48)}\right]$$

$$\text{BLYAS bioluminescence} = 456.1 + (9037.8 - 456.1)/\left[1 + \left([\text{DHT}]/(4.2 \times 10^{-9})\right)^{(-1.76)}\right]$$

3. To calculate the equivalent E2 or DHT concentration in the sample, use only data points from sample concentrations that are not toxic (i.e., relative bioluminescence >0.8 as determined in the BLYR assay) and result in a bioluminescent response greater than mean + 3× standard deviation of the methanol controls in the BLYES and BLYAS assays. In this example, samples at 25×, 12.5×, and 5× exposure concentration meets this criteria in the BLYES assay and therefore are used to calculate the equivalent E2 concentration of the preconcentrated (1×) sample, resulting in 2.5×10^{-12} M, 4.3×10^{-12} M, and 6.8×10^{-12} M, respectively. The equivalent E2 concentration of the 1× sample is then reported as the average of the three calculated values (3.4×10^{-12} M). Similarly, the equivalent DHT concentrations of the 1× sample is calculated to be 2.1×10^{-11} M using bioluminescence data from BLYAS cells exposed to 50× and 25× sample concentrations.

4 Notes

1. Plasticware should not be used for medium storage and yeast growth as they can leach estrogenic and/or androgenic chemicals during use, leading to false positive results. Baking in a muffle furnace is highly recommended for glassware sterilization to remove any organic residue that might interfere with the assay.
2. The DHT standards are prepared at a different concentration range from the E2 standards due to different sensitivities between the BLYAS and BLYES strains.
3. Repeated liquid subculture is not recommended and can lead to decreased sensitivity. To make frozen stocks, inoculate an isolate grown on YMM agar without leucine, uracil, or tryptophan into 30 mL fresh liquid medium in a baked sterile glass flask and incubate at 28 °C with 200 rpm shaking to an OD_{600} of 1.0. Mix the yeast culture with an equal volume of 50%

sterile glycerol (prepared with HPLC water and stored in a baked sterile glass bottle). Aliquot the yeast–glycerol mixture into individual cryogenic tubes. Store at −80 °C.

4. The BLYES and BLYAS strains produce optimal sensitivity to target chemical at this range of cell density and growth phase. Assays with a lower cell density may result in a delay in response time, whereas reduced sensitivity may be observed when using a stationary phase culture.
5. It is necessary to use the same aliquot of methanol that is used for standard and sample preparation as negative controls.
6. The methanol negative controls and yeast blank controls should produce similar bioluminescence. An increase in bioluminescence in the methanol controls indicates contamination in the solvent.
7. The integration time can be adjusted from 0.1 to 1 s/well depending on instrument sensitivity. The reading intervals and total assay time can also be adjusted based on specific applications. Assays longer than 8 h are not recommended due to decreased yeast activity.
8. Induction of bioluminescence in response to E2 or DHT exposure can be detected in as early as 3 h. However, 4–8 h exposure is recommended to achieve optimal sensitivity.

Acknowledgments

Research funding was provided by the National Institute of Food and Agriculture, US Department of Agriculture, under award number 2015-33610-23598 and the US National Science Foundation, Chemical, Bioengineering, Environmental and Transport Systems (CBET) program, under award number 1530953.

References

1. Vogel J (2004) Tunnel vision: the regulation of endocrine disruptors. Pol Sci 37(3):277–303
2. Kabir ER, Rahman MS, Rahman I (2015) A review on endocrine disruptors and their possible impacts on human health. Environ Toxicol Pharmacol 40(1):241–258. https://doi.org/10.1016/j.etap.2015.06.009
3. Sanseverino J, Gupta RK, Layton AC, Patterson SS, Ripp SA, Saidak L, Simpson ML, Schultz TW, Sayler GS (2005) Use of *Saccharomyces cerevisiae* BLYES expressing bacterial bioluminescence for rapid, sensitive detection of estrogenic compounds. Appl Environ Microbiol 71(8):4455–4460. https://doi.org/10.1128/AEM.71.8.4455-4460.2005
4. Eldridge M, Sanseverino J, Layton A, Easter J, Schultz T, Sayler G (2007) *Saccharomyces cerevisiae* BLYAS, a new bioluminescent bioreporter for detection of androgenic compounds. Appl Environ Microbiol 73(19):6012–6018
5. Gupta RK, Patterson SS, Ripp SA, Sayler GS (2003) Expression of the *Photorhabdus luminescens lux* genes (*luxA*, *B*, *C*, *D*, and *E*) in *Saccharomyces cerevisiae*. FEMS Yeast Res 4(3):305–313
6. Sanseverino J, Eldridge ML, Layton AC, Easter JP, Yarbrough J, Schultz TW, Sayler GS (2009) Screening of potentially hormonally active chemicals using bioluminescent yeast bioreporters.

Toxicol Sci 107(1):122–134. https://doi.org/10.1093/toxsci/kfn229

7. Krifaton C, Kriszt B, Risa A, Szoboszlay S, Cserháti M, Harkai P, Eldridge M, Wang J, Kukolya J (2013) Application of a yeast estrogen reporter system for screening zearalenone degrading microbes. J Hazard Mater 244:429–435. https://doi.org/10.1016/j.jhazmat.2012.11.063
8. Osimitz T, Eldridge M, Sloter E, Welsh W, Ai N, Sayler G, Menn F, Toole C (2012) Lack of androgenicity and estrogenicity of the three monomers used in Eastman's Tritan copolyesters. Food Chem Toxicol 50(6):2196–2205
9. Balazs A, Krifaton C, Risa A, Cserhati M, Kukolya J, Toth A, Hahn J, Szoboszlay S, Eldridge M, Wang J, Kriszt B (2014) Biodegradation of 5 alpha-dihydrotestosterone to non-androgenic products. Int Biodeterior Biodegradation 93:162–167. https://doi.org/10.1016/j.ibiod.2014.05.008
10. Bergamasco AMD, Eldridge M, Sanseverino J, Sodre FF, Montagner CC, Pescara IC, Jardim WF, Umbuzeiro GD (2011) Bioluminescent yeast estrogen assay (BLYES) as a sensitive tool to monitor surface and drinking water for estrogenicity. J Environ Monit 13(11):3288–3293. https://doi.org/10.1039/c1em10464k
11. Jardim W, Montagner C, Pescara I, Umbuzeiro G, Bergamasco A, Eldridge M, Sodre F (2012) An integrated approach to evaluate emerging contaminants in drinking water. Sep Purif Technol 84:3–8. https://doi.org/10.1016/j.seppur.2011.06.020
12. Ruan T, Liang D, Song SJ, Song MY, Wang HL, Jiang GB (2015) Evaluation of the *in vitro* estrogenicity of emerging bisphenol analogs and their respective estrogenic contributions in municipal sewage sludge in China. Chemosphere 124:150–155. https://doi.org/10.1016/j.chemosphere.2014.12.017
13. Wang J, Eldridge M, Menn FM, Dykes T, Sayler G (2015) Standardized application of yeast bioluminescent reporters as endocrine disruptor screen for comparative analysis of wastewater effluents from membrane bioreactor and traditional activated sludge. Ecotoxicology 24(10):2088–2099. https://doi.org/10.1007/s10646-015-1556-z

Chapter 4

Bioluminescent Imaging of Single Bacterial Cells Using an Enhanced *ilux* Operon

Carola Gregor

Abstract

The *lux* operon is a useful reporter for bioluminescence imaging due to its independence of exogenous luciferin supply, but its relatively low brightness hampers the imaging of single cells. This chapter describes a procedure for the imaging of individual *Escherichia coli* cells using an improved *ilux* operon. The enhanced brightness of *ilux* enables long-term bioluminescence imaging of single bacteria with high sensitivity without the requirement for an external luciferin.

Key words Bioluminescence, Luciferase, Bacteria, Microscopy, Imaging, *lux*

1 Introduction

Bioluminescence imaging is an important technique for a wide range of different applications, but it often suffers from the requirement of luciferin addition and its relatively low light levels compared to fluorescence measurements. Up to now, bacterial bioluminescence has the unique feature that all enzymes involved in synthesis and recycling of the luciferin have been identified.[1] This means that the corresponding genes can be coexpressed together with the luciferase to enable continuous bioluminescence emission without the addition of any substrates. The genes required for the generation of bacterial bioluminescence are coded in the *lux* operon, which can also be expressed and used as a reporter in nonbioluminescent bacteria. For instance, bacterial bioluminescence has been used for in vivo imaging of bacterial infections [2–4] and tumors [5, 6], as well as for assaying gene expression [7–12] and toxicity [13–16]. However, the brightness of bacterial bioluminescence emission is comparatively low, which is an obstacle for single-cell imaging applications.

[1]Very recently, Kotlobay et al. described the identification of the enzymes of the fungal bioluminescence system [1]. This makes the fungal system now the second bioluminescence system that is fully genetically encodable.

Steven Ripp (ed.), *Bioluminescent Imaging: Methods and Protocols*, Methods in Molecular Biology, vol. 2081,
https://doi.org/10.1007/978-1-4939-9940-8_4,

In order to overcome the limitation of its low brightness, we have engineered an improved *lux* operon named *ilux* [17]. This operon consists of the *luxCDABE* genes from *Photorhabdus luminescens* and the *frp* gene from *Vibrio campbellii* for increased $FMNH_2$ production. Additionally, all genes of the operon were varied by multiple rounds of error-prone mutagenesis and screened for enhanced brightness, resulting in several mutations in the *luxA*, *luxB*, *luxC*, and *frp* genes. *ilux* produces ~7-fold greater bioluminescence in *Escherichia coli* cells compared to the wild-type *luxCDABE* operon, with an optimum temperature of 37 °C. Due to their increased light emission, single *ilux*-expressing cells can be imaged by bioluminescence microscopy with high signal-to-noise ratio. For this purpose, the resuspended bacteria are embedded between a coverslip and an agar pad to avoid movement of the cells. This procedure allows for continuous imaging over time periods of up to several days without any further treatment of the sample.

2 Materials

2.1 Microscope

Figure 1 shows the schematic construction of our custom-built microscope which consists of the following components:

1. Optical table (M-ST-UT2-410-18, Newport) with pneumatic vibration isolators (I-2000, Newport).
2. Water-cooled EMCCD (electron multiplying charge-coupled device) camera, 128 × 128 pixels (iXon DU860, Andor).

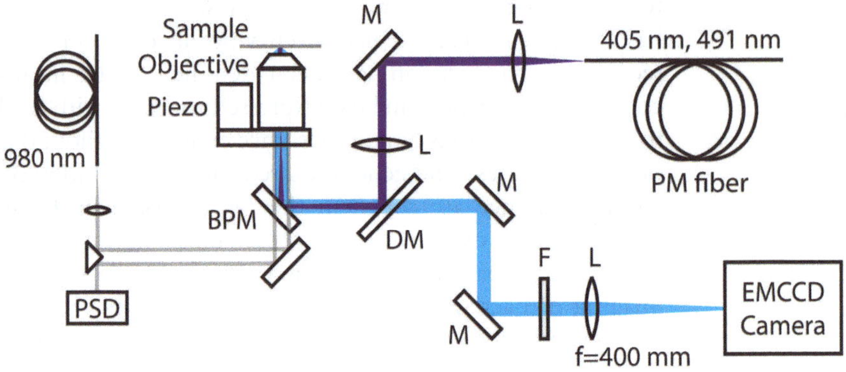

Fig. 1 Schematic setup of the microscope. Bioluminescence light is collected by an oil immersion objective lens and directed to an EMCCD camera using dielectric mirrors (back side-polished dielectric mirror (BPM) and mirror (M)) optimized for the visible wavelength range. The light is spectrally filtered by a short-pass filter (F) and focused onto the camera using a lens (L) with a focal length of 400 mm. For focusing and selection of cells, a wide-field excitation with 405 nm and 491 nm is implemented. Both lasers are coupled into the setup through a polarization-maintaining (PM) optical fiber. A dichroic mirror (DM) is used to separate the excitation light from fluorescence. Long-term stability of the focus position is provided by a custom-built focus lock system. This is based on the detection of a TIR signal using a position-sensitive diode (PSD) and repositioning the objective lens with a z piezo. In- and outcoupling of the 980 nm light for the focus lock is achieved through a BPM. (Reproduced from [17])

3. Oil immersion objective lens (HC PL APO 100×/1.40–0.70 OIL CS, Leica).
4. Custom-built manual microscope stage.
5. Objective positioner piezo (MIPOS 100 PL CAP, piezosystem Jena).
6. 405 nm laser for excitation of *ilux* fluorescence (PhoxX 405–60, Omicron or DL 100, Toptica).
7. 491 nm laser for additional fluorescence measurements (Calypso 50 mW, Cobolt).
8. 980 nm laser for focus lock (SPL-980-15-9-PD, Roithner Lasertechnik).
9. Dichroitic beam splitter (ZT405rdc, Chroma).
10. Fluorescence emission filters: movable filter HQ 520/40M, Chroma (for 491 nm laser), BrightLine FF01-842/SP, Semrock (for 980 nm laser).
11. Dielectric mirrors (BB1-E02 and BB1-E02P, Thorlabs).
12. Achromatic lens (VIS coated, focal length 400 mm, Qioptiq).
13. Position-sensitive diode for focus lock system (DL-100-7-KER pin, Silicon Sensor).
14. Two custom-made black boxes around the optical components on the optical table.
15. Measurement control software, custom-written in Python.

2.2 Expression of ilux

1. *ilux*-containing expression plasmid (e.g., *ilux* pGEX(−) (addgene plasmid #107879) or *ilux* pQE(−)). Both plasmids are available upon request from the author.
2. Competent bacteria.
3. LB medium: 5 g/l yeast extract, 5 g/l NaCl, 10 g/l peptone, 5 mM NaOH. Alternatively, a different medium may be used.
4. Agar plates: LB medium (or other medium of choice) containing 14 g/l agar and appropriate antibiotic (50 mg/l ampicillin for *ilux* pGEX(−) and *ilux* pQE(−)), optionally IPTG (around 20 μM).
5. Incubator.

2.3 Sample Preparation

1. Sample chamber (Fig. 2, *see* **Note 1**).
2. Coverslips (18 mm diameter).
3. Glue for mounting the coverslip into the sample chamber (e.g., picodent twinsil speed 22, picodent).
4. Colonies of *ilux*-expressing bacteria.
5. LB medium: 5 g/l yeast extract, 5 g/l NaCl, 10 g/l peptone, 5 mM NaOH. Alternatively, a different medium may be used.

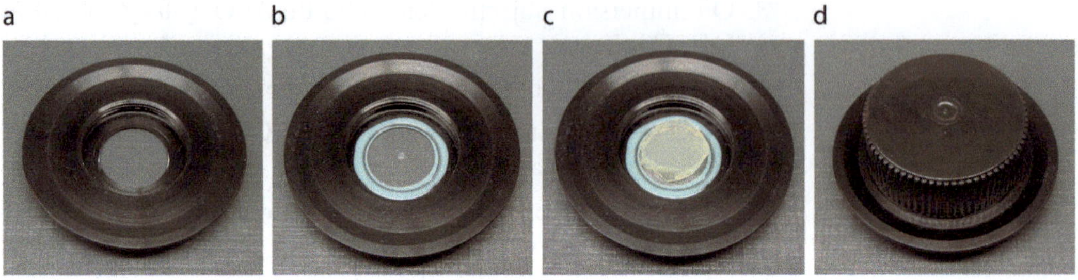

Fig. 2 Sample preparation for imaging. (**a**) Empty custom-built sample chamber. (**b**) A coverslip (18 mm diameter) is glued into the sample chamber and a drop of *E. coli* suspension is placed in its center. (**c**) The cells are covered with an agar pad. (**d**) A blackened lid from a 50 ml Falcon tube is put on top of the sample

6. Agar plates: LB medium (or other medium of choice) containing 14 g/l agar and appropriate antibiotic (50 mg/l ampicillin for *ilux* pGEX(−) and *ilux* pQE(−)), optionally IPTG (around 20 μM).
7. Scalpel.

3 Methods

3.1 Construction of the Microscope

Details of construction of a custom-built microscope are described in [17]. For the design of the setup, the following issues should be considered:

1. In order to detect as much of the bioluminescence light as possible, use an objective lens with a high numerical aperture. Additionally, the absorbance of all optical components (e.g., filters for fluorescence measurements) should be as low as possible in the spectral window of bioluminescence emission (~400–600 nm).
2. An EMCCD camera should be used to obtain a high sensitivity of detection. Cool the camera sensor to the lowest possible temperature to reduce the dark noise. In addition, adapt the effective pixel size on the camera to a value as large as possible for the required image resolution. As a result, the emitted bioluminescence light will be distributed over fewer pixels and the signal-to-noise ratio (SNR) is improved.
3. Implement a light source to select and focus the cells, for instance a 405 nm laser (*see* **Note 2**), and additional excitation lasers (illumination by a suitable lamp may be substituted) for fluorescence measurements if needed.
4. For long-term measurements, a focus lock system is advantageous to keep the cells in focus during imaging.
5. To obtain maximum levels of bioluminescence, heating may be used. In *E. coli* cells, heating the sample to 37 °C increases the

brightness by a factor of 2–3 compared to room temperature. For other bacteria, different temperatures around their optimum growth temperature may be tested for best results.

6. Take great care to shield the microscope from stray light from the surroundings to keep the background signal as low as possible. For this purpose, 2–3 black boxes (e.g., made of black cardboard) may be built around the setup. If the setup is not sufficiently protected from ambient light, measurements will have to be performed in a fully darkened room.

3.2 Expression of ilux

The following protocol is optimized for *E. coli* cells. For bacteria other than *E. coli*, medium and expression conditions may need to be adapted.

1. Clone the *ilux* operon into a suitable expression vector. For expression in *E. coli*, two plasmids are available from the author or addgene: *ilux* pGEX(−) (addgene #107879) and *ilux* pQE (−) (*see* **Note 3**). For bacteria other than *E. coli*, choose an appropriate expression vector with a strong promoter and without a tag (*see* **Notes 4** and **5**).
2. Transform your bacteria of interest with the *ilux*-containing expression plasmid. Spread different dilutions on LB agar plates containing the appropriate antibiotic. Let the bacteria grow overnight at a suitable temperature (37 °C for *E. coli*).
3. The next day, colonies should have appeared on the agar plates. Visually inspect the plates for bioluminescence emission in a dark room. Bioluminescence light should be easily visible with the naked eye. Otherwise, the expression may need to be optimized for successful single-cell imaging.

3.3 Sample Preparation

1. Prepare the sample chamber (*see* **Note 1**) by gluing a coverslip at the bottom (Fig. 2). Let the glue completely harden before proceeding to the next step to avoid drifting of the sample during imaging.
2. Scrape *ilux*-expressing cells off the agar plate with a sterile pipette tip and resuspend them in 10 μl of water, buffer or medium by pipetting or vortexing. Resuspend thoroughly in order to separate the cells from each other (*see* **Note 6**). Use the cells immediately for imaging.
3. Place 0.5–1 μl of the cell suspension at the center of the coverslip.
4. Cut a round agar pad with a diameter of 1.0–1.5 cm from an LB agar plate. Put the agar pad onto the cells upside down (*see* **Note 7**). Do not shift the agar pad. Slightly press the agar pad onto the coverslip.
5. Optional: cover the sample chamber with a lid (e.g., the lid of a 50 ml Falcon tube, *see* **Note 8**).

3.4 Imaging

1. Before starting the measurements, make sure that light from the surroundings is efficiently excluded from the microscopy setup. If you are not sure whether a certain light source (e.g., computer screen or room light) will disturb your measurements, take test images without a sample where the light source is switched on and off, respectively. Both images should exhibit identical levels of homogeneously distributed background signal.
2. Mount the sample chamber onto the microscope stage. Focus the cells (e.g., using their fluorescence signal, *see* **Note 2**) and select a field of view with an appropriate number of isolated cells (*see* **Note 9**).
3. If using a focus lock system, switch on the focus lock (*see* **Note 10**). Switch off the light source utilized for focusing and all other lasers if present.
4. Start the bioluminescence image acquisition. Depending on the brightness of the sample, adjust the exposure time (*see* **Note 11**). Adjust the camera settings (e.g., readout speed, preamplification gain, and electron magnification gain) to obtain an optimal SNR.

3.5 Data Postprocessing

1. If images contain bright spots originating from cosmic rays (Fig. 3), these can be filtered out by applying a simple brightness threshold if their brightness is considerably higher than the bioluminescence signal. If the bioluminescence signal is similarly high, the brightness of each pixel may be compared with its value in the previous and/or following image in a time series, and a threshold applied to their difference. Alternatively, the neighboring pixels may be used for comparison if no consecutive images are available.

   The corresponding pixel values may for instance be replaced by the average value of the same pixel in the previous and following image, or by an average value of the surrounding pixels.
2. From each pixel, subtract the average pixel value of an image taken with the same exposure time without a sample if desired. Set all resulting negative pixel values to zero.
3. Scale the colormap between the minimum and maximum pixel value of the image or image series in order to obtain the best image contrast. For example, in a gray-scale image the darkest pixel should be depicted in black and the brightest pixel in white.

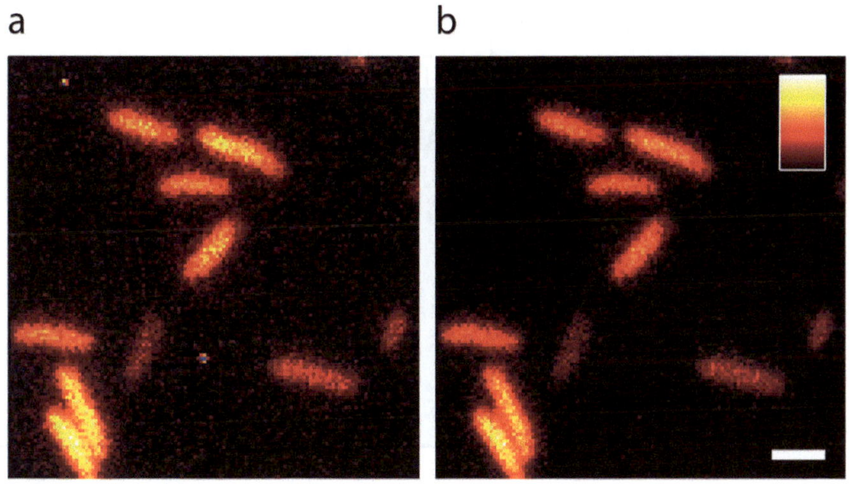

Fig. 3 Sample image from a time series of *E. coli* Top10 cells expressing *ilux* from the vector pQE(−). Cells were imaged under an LB agar pad at room temperature with an exposure time of 10 min. (**a**) Raw image containing two bright spots due to cosmic rays. Blue pixels represent saturation. (**b**) Processed image after subtraction of the average background signal of a measurement without a sample. In addition, pixels with a brightness higher than the value of the same pixel in the following image plus a threshold value were replaced by the average value of the same pixel in the previous and following image in order to remove the bright spots. The colormap was scaled to the minimum and maximum pixel values of each image (excluding the bright spots). Scale bar: 2 μm

4 Notes

1. A sample chamber compatible with the microscope stage is required (a home-built sample chamber is shown in Fig. 2). Since sufficient oxygen supply is important for high levels of bioluminescence, the cells should not be embedded between an object slide and coverslip, but imaged in an open chamber.
2. *E. coli* cells expressing the *ilux* proteins exhibit relatively bright fluorescence when excited with 405 nm light, which possibly originates from the luciferase-bound C4a-peroxyflavin or C4a-hydroxyflavin intermediate of the bioluminescence reaction. This fluorescence emission can be used to select and focus the cells before starting the bioluminescence measurement. However, since the fluorescence bleaches very rapidly, it is crucial to keep both the exposure time and the intensity of the excitation light as low as possible. Alternatively, other techniques such as phase-contrast microscopy may be used to visualize the cells.
3. In *E. coli* Top10 cells, expression of *ilux* from pQE(−) produces higher levels of bioluminescence than from pGEX(−) without the need for IPTG induction (Fig. 4). The brightness of uninduced *ilux* pGEX(−) is ~10-fold lower, but increases to 75% of the brightness of *ilux* pQE(−) if 20 μM IPTG is added

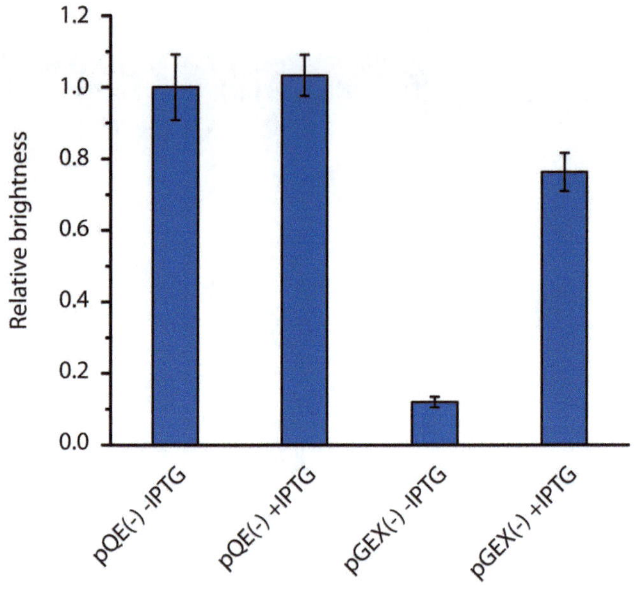

Fig. 4 Relative brightness of *ilux* expressed from pGEX(−) and pQE(−) in *E. coli* Top10 cells on LB agar plates. Single colonies were spread out onto a new agar plate with or without 20 μM IPTG and grown overnight and imaged at 37 °C. Error bars represent standard deviation of 10 different clones

to the medium. However, *E. coli* DH5α cells with *ilux* pQE(−) do not grow. Depending on the bacterial strain, it may be beneficial to exchange the expression vector and to induce the expression if necessary.

4. Since the brightness of bioluminescence emission depends on the concentrations of three reaction partners whose formation is driven by the *ilux* operon (luciferase, $FMNH_2$ and aldehyde), it strongly depends on the expression levels. Therefore, it is important to select an expression vector with a strong promoter and high copy number in order to maximize the brightness. On the other hand, excessive expression levels may impede cell growth due to accumulation of the toxic aldehyde substrate and should hence be avoided.

5. Fusion tags either at the N-terminus of *luxC* or at the C-terminus of *frp* are likely to reduce the brightness significantly and should therefore be omitted. If no tag-free expression vector is available, a vector with a C-terminal tag may be used if a stop codon is included at the end of the *frp* gene. Alternatively, remove the tag from the vector before cloning of the *ilux* insert.

6. If the cells are not thoroughly resuspended, cell aggregates will be observed. Cells may be difficult to resuspend if the plates with the colonies are kept at 4 °C for long periods of time.

7. If the cells are observed to drift during imaging, a liquid film may have formed between the coverslip and the agar. In that case, prepare a new sample with a smaller volume of cell suspension and let the agar pad air-dry for 5–30 min at room temperature before putting it onto the cells.
8. Covering the sample chamber will prevent drying out of the agar during long-term imaging and may additionally help to absorb stray light. The lid should be put only loosely onto the chamber not to hamper oxygen supply.
9. If clusters of cells are observed or if the cell density is too high or too low, prepare a new sample. Too high a cell density will increase the background signal due to scattered bioluminescence light and should therefore be avoided. This will even be observed if the cell density is low within the field of view, but high in the surrounding area.
10. If a focus lock system based on total internal reflection at the coverslip-agar interface as described in [17] is used, the lock will not function properly if the interface is disturbed by a cell layer formed by dividing cells. In that case, the focus lock should be switched off before the cells become too dense.
11. Several images may be averaged during postprocessing of the data to increase the SNR. However, due to repeated read noise this is not recommended if a higher SNR can instead be obtained by prolonging the exposure time.

Acknowledgments

I thank Prof. Stefan W. Hell for the opportunity to carry out this work in his department.

References

1. Kotlobay AA et al (2018) Genetically encodable bioluminescent system from fungi. Proc Natl Acad Sci U S A 115(50):12728–12732
2. Karsi A, Menanteau-Ledouble S, Lawrence ML (2006) Development of bioluminescent *Edwardsiella ictaluri* for noninvasive disease monitoring. FEMS Microbiol Lett 260:216–223
3. Gonzalez RJ, Weening EH, Frothingham R, Sempowski GD, Miller VL (2012) Bioluminescence imaging to track bacterial dissemination of *Yersinia pestis* using different routes of infection in mice. BMC Microbiol 12:147
4. Huang YK, Chu C, Wu CH, Chen CL, Chiu CH (2014) Evaluation of Gram-negative bacterial infection by a stable and conjugative bioluminescence plasmid in a mouse model. J Biomed Sci 21:78
5. Nguyen VH, Kim HS, Ha JM, Hong Y, Choy HE et al (2010) Genetically engineered *Salmonella typhimurium* as an imageable therapeutic probe for cancer. Cancer Res 70:18–23
6. Cronin M, Akin AR, Collins SA, Meganck J, Kim JB et al (2012) High resolution *in vivo* bioluminescent imaging for the study of bacterial tumour targeting. PLoS One 7: e30940
7. Engelbrecht J, Simon M, Silverman M (1985) Measuring gene expression with light. Science 227:1345–1347
8. Shen H, Gold SE, Tamaki SJ, Keen NT (1992) Construction of a Tn7-*lux* system for gene

expression studies in gram-negative bacteria. Gene 122:27–34
9. Kondo T, Strayer CA, Kulkarni RD, Taylor W, Ishiura M et al (1993) Circadian rhythms in prokaryotes: luciferase as a reporter of circadian gene expression in cyanobacteria. Proc Natl Acad Sci U S A 90:5672–5676
10. Sayler GS, Fleming JT, Nivens DE (2001) Gene expression monitoring in soils by mRNA analysis and gene *lux* fusions. Curr Opin Biotechnol 12:455–460
11. Shimizu T, Ohta Y, Tsutsuki H, Noda M (2011) Construction of a novel bioluminescent reporter system for investigating Shiga toxin expression of enterohemorrhagic *Escherichia coli*. Gene 478:1–10
12. Vannini A, Agriesti F, Mosca F, Roncarati D, Scarlato V et al (2012) A convenient and robust *in vivo* reporter system to monitor gene expression in the human pathogen *Helicobacter pylori*. Appl Environ Microbiol 78:6524–6533
13. Ulitzur S, Lahav T, Ulitzur N (2002) A novel and sensitive test for rapid determination of water toxicity. Environ Toxicol 17:291–296
14. Min J, Chang YS, Gu MB (2003) Bacterial detection of the toxicity of dioxins, polychlorinated diphenyls, and polybrominated diphenyl ethers. Environ Toxicol Chem 22:2238–2242
15. Kelly CJ, Tumsaroj N, Lajoie CA (2004) Assessing wastewater metal toxicity with bacterial bioluminescence in a bench-scale wastewater treatment system. Water Res 38:423–431
16. Ivask A, Rõlova T, Kahru A (2009) A suite of recombinant luminescent bacterial strains for the quantification of bioavailable heavy metals and toxicity testing. BMC Biotechnol 9:41
17. Gregor C, Gwosch KC, Sahl SJ, Hell SW (2018) Strongly enhanced bacterial bioluminescence with the *ilux* operon for single-cell imaging. Proc Natl Acad Sci U S A 115:962–967

Chapter 5

Bioluminescent Imaging and Tracking of Bacterial Transport in Soils

Jie Zhuang, Weipeng Liu, Liqiong Yang, Jia Kang, and Xiaoming Zhang

Abstract

Bioimaging instrumentation can be used to observe environmental phenomena such as the transport, retention, and distribution of bacteria in soils in situ in a real-time, nondestructive manner. Bacteria designed to emit bioluminescence light signals are injected into a transparent column packed with soils, and then the column is placed into a bioimaging instrument, such as a PerkinElmer IVIS Spectrum, while it is connected through thin teflon tubes to other parts of the column system located outside of the imaging chamber, including a fraction collector for collecting effluent solution and a pump for introducing bacterial suspension or experimental solution. After self-correction of soil autofluorescence and bioluminescence and setup of required imaging parameters, the transport experiment is initiated by introducing the bacterial suspension to the soil column while the spatiotemporal distribution of bioluminescent bacteria in the entire soil column is imaged. Finally, the images are processed to analyze bacterial migration in the soil under various environmental conditions in comparison with the breakthrough and elution curves of the bacteria obtained by analyzing the effluent samples.

Key words Bioluminescence, *lux*, Soil heterogeneity, Bacterial transport, Profile distribution of bacteria

1 Introduction

Contamination by pathogenic microorganisms is a frequent occurrence that threatens public drinking water, agricultural irrigation, animal husbandry, and food production. Mechanistic understanding of pathogen transport in natural soils, particularly in undisturbed soils, is a challenge. Advances in this area will greatly benefit the assessment and prediction of pathogen transport in subsurface environments and assist in the development of management strategies for water safety and cleanliness relevant to pathogen contamination. In the past decades, great advances have been made in quantifying the fate and transport of microorganisms in soils. However, most studies have been conducted by taking the soil as a black or grey system, lacking in situ information on the real-time

Steven Ripp (ed.), *Bioluminescent Imaging: Methods and Protocols*, Methods in Molecular Biology, vol. 2081,
https://doi.org/10.1007/978-1-4939-9940-8_5,

processes within soil structure or between soil layers. The major obstacles for mechanistic understanding and accurate assessment on the effect of soil heterogeneity on microbial transport are the lack of knowledge in spatial variations of transport processes. It is thus extremely important to develop in situ noninvasive visualization methods for the quantitative observation of the spatially varying distribution and transport of pathogens in soils in a real-time manner.

A number of spectrum and spectral imaging techniques have been developed to count microorganisms and analyze their properties. Infrared spectroscopy (IRS) can identify microorganisms according to the distinct infrared spectra provided by unique chemical components in their cell membranes and cell walls. In the past decade, IRS has been widely used for the discrimination and the classification of various microorganisms at the levels of species, subspecies, strains, serotype, and haplotype based on the infrared absorbance patterns of cell components [1]. Surface-enhanced Raman spectroscopy (SERS) technology reflects the structure and chemical information of samples through enhanced Raman scattering effects. It can be used not only to conduct qualitative and quantitative analysis of individual compounds but also to detect microorganisms [2]. The terahertz (THz) radiation can excite the low-frequency biomolecular (such as protein and DNA) motions, and thereby it is available for acquiring unique structural and dynamic information that is absent in other electromagnetic spectroscopies [3]. THz-based spectroscopic sensing and imaging techniques have rapidly developed for Terahertz time-domain spectroscopy (THz-TDS), which is widely used for microbial survival assessment [4]. Laser induced breakdown spectroscopy (LIBS) is a relatively new and versatile spectroscopic technique for in situ elemental detection and quantitative chemical analysis based on atomic emission spectroscopy. LIBS technology has been used for bacterial identification and differentiation since its initial demonstration in 2003 [5, 6]. As for spectral imaging techniques, hyperspectral imaging (HSI), also known as imaging spectrometry, integrates both imaging or computer vision and spectroscopic techniques into one system. In comparison with routine imaging or spectroscopic techniques, HSI can measure both spatial information and spectral parameters for each pixel in the image simultaneously [7]. In terms of HSI, several systems, including visible (Vis), near-infrared, FTIR, fluorescence, and Raman HIS, are available and highly flexible. Hyperspectral images have been used to detect and quantify microorganism distribution in real samples [8]. Multispectral imaging (MSI) is another spectral imaging technology based on HSI technology. In general, the instrument complexity, cost, and data acquisition time of MSI systems are significantly lower than those of HSI systems. Recently, MSI systems combined with chemometrics have been employed for microorganism detection in meat products [9], and the detection is very effective.

The IVIS imaging systems designed by PerkinElmer, as well as other similar bioimaging instruments marketed by Spectral Instruments Imaging, Bruker, Biospace Lab, Li-Cor, Berthold Technologies, and MI Labs, provide high-sensitivity, low-noise, in vivo imaging platforms that enable noninvasive visualization and tracking of cellular and genetic activity within a living organism in real time. These systems typically provide both bioluminescence and fluorescence imaging capabilities. Bioluminescence imaging sensitivity is usually 1000 times higher than fluorescence imaging due to the lack of autobioluminescence background from most materials. In our study, we use bioluminescence imaging to acquire real-time images of microorganisms in soils for tracking their real-time movement, retention, detachment, and distribution in soil under a variety of flow and soil conditions.

2 Materials

All solutions are prepared using ultrapure water, which is produced by purifying deionized water to attain a sensitivity of 18 MΩ-cm at 25 °C.

2.1 Bacterial Strains

Pseudomonas fluorescens 5RL (*Pf*5RL): Contains the *lux* operon, was obtained from the University of Tennessee's Center for Environmental Biotechnology in Knoxville, Tennessee, USA. Other bioluminescent bacteria can be obtained through microbial repositories or personal requests (*see* **Note 1**). Growth methods provided here are for *P. fluorescens* and will need to be modified if other bacterial strains are used.

2.2 Chemicals

1. Luria–Bertani (LB) media: 10 g tryptone, 10 g NaCl, and 5 g yeast extract in 1 L H_2O, with pH adjusted to 7.0 using 0.5 M NaOH.
2. Sodium chloride background solution.
3. Tetracycline (CAS 60-54-8)
4. NaBr.
5. Naphthalene.

2.3 Porous Materials

Porous materials: clean sand, humic acid (HA)-coated sand, goethite-coated sand, or sieved soil. Different particle sizes and treatments of sand and soil have different effects on the bioluminescence imaging of bacterial solution (*see* **Notes 2–4**) (Fig. 1).

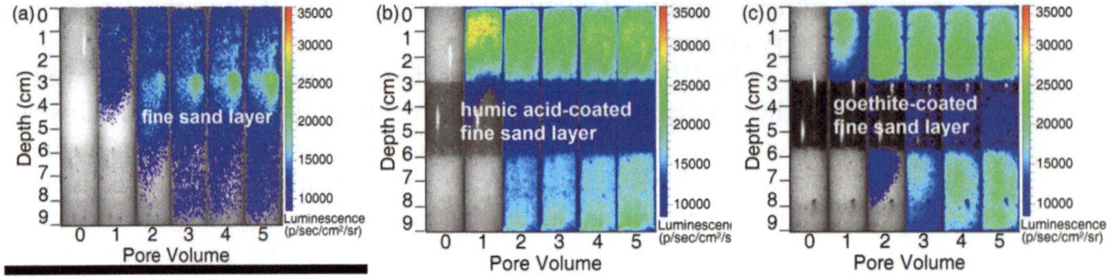

Fig. 1 Real-time distribution of bacteria (*Pf*5RL) at different pore volumes during downward transport (**a**) in coarse sand with a horizontal layer of fine sand, (**b**) in coarse sand with a horizontal layer of humic acid-coated fine sand, and (**c**) in coarse sand with a horizontal layer of hematite-coated fine sand. Results show that the coatings of humic acid and goethite affect bioluminescence imaging

2.4 Imaging Instrumentation

These experiments used a PerkinElmer IVIS Spectrum Imaging System. Other comparable imaging systems can be similarly used.

2.5 Other Supplies and Equipments

1. Shake flasks.
2. Rotary shaker.
3. Spectrophotometer.
4. Centrifuge.
5. Opaque 96-well microtiter plate.
6. Acrylic or other transparent column.
7. Nylon membranes or glass fiber.
8. Ion chromatograph.

3 Methods

3.1 Preparation of the Bacterial Suspension

1. Sterilize the shake flask following autoclave procedures: 30 min, 121 °C.
2. Transfer strain to a sterile 250-mL baffled shake flask containing 100 mL of LB. Add kanamycin at a concentration of 10 mg/L to the *P. fluorescens* to ensure that only the kanamycin resistant strains grow.
3. Culture cells in a rotary shaker at 160 rpm and 30 °C.
4. Sodium chloride background solution: For different ionic strengths, make decisions based on your own experimental needs.
5. The culture was grown to stationary phase (optical density of 1.0 at a wavelength of 600 nm), centrifuged at 1000 × *g* for 10 min, and resuspended in NaCl solution.

3.2 Column System Treatment

Acrylic or other transparent column (*see* **Note 5**) of appropriate length to fit into instrument's imaging chamber. The diameter should be as small as possible without affecting the experiment.

1. The porous materials and columns are autoclaved to sterilize (*see* **Note 6**).
2. The porous materials are dry packed into a transparent column in 1-cm increments with stirring and tapping (*see* **Note 7**).
3. Place nylon membranes or glass fiber at the column inlet and outlet to prevent the porous materials from being flushed out.
4. After packing, high pressure CO_2 gas is introduced upward into the column at a low flow rate for at least 3 h to displace trapped air.
5. The magnitude of a pore volume (PV):

$$\mathrm{PV} = V - \frac{m}{\mathrm{sp\ gr}}$$

   where V is the internal volume of the columns; m is the weight of the porous materials; sp gr is the specific gravity of the porous materials. The number of PV is used for controlling experimental time and processing of breakthrough and elution results.
6. The packed column is flushed with sterile background solution (NaCl) at a constant pore velocity for at least 20 pore volumes.
7. The bacterial suspension is pumped into the column after the column system is stabilized in both hydrodynamics and chemistry.
8. The input bacterial concentration is ~10^8 cells/mL, containing 30 mg/L NaBr (*see* **Note 8**).
9. Effluent samples are collected in every pore volume.
10. The column is flushed with background solution immediately after collecting a given number of pore volumes of effluent samples. Samples are collected per pore volume.
11. Once the experiment is terminated, 1-cm segments of porous media are taken from the column to examine the retention profile of bacteria along the travel pathway (*see* **Note 9**).
12. Examine the detachment of retained bacteria in the column and estimate the total mass recovery of injected bacteria:

$$M_{\mathrm{tot}} = M_{\mathrm{eff}} + M_{\mathrm{elu}} + M_{\mathrm{ret}}$$

   where M_{tot} is the total mass recovery, M_{eff} is the effluent mass recovery, M_{elu} is the elution mass recovery, and M_{ret} is the retention mass recovery.

3.3 Flat Colony Counting Method

1. Dilute solution: normal saline or other.
2. Sterilize glassware and normal saline.

3. Under aseptic operation, add the bacteria solution into the glass bottle and shake well to make 1:10 uniform diluents. 1 mL of 1:10 diluent is absorbed and slowly injected into the test tube containing 9 mL of sterilized normal saline or other diluents along the wall of the tube. The shaking test tube is mixed evenly to make 1:100 diluent. According to the above operation sequence, make ten times increasing diluent with new 1 mL of sterilization straw.
4. Select appropriate dilution, add 1 mL of the diluent to the petri dish as control, at a medium temperature of about 45 °C.
5. After the medium is solidified, flip the plate and place it in a constant temperature incubator for culture: 30 °C, 24 h.
6. After bacterial culture, the plates with colonies between 30 and 300 are selected for counting, and the number of colonies in the original sample is calculated according to the dilution factor.

3.4 Breakthrough Curves of Bromide

Bromide (30 mg/L in NaBr) is used as a conservative tracer for quantifying the dispersivity and hydrodynamic conditions of the flow-through experiments. Effluent samples (from Subheading 3.2, **steps 9** and **10**) are used to analyze Br^- concentration using Ion Chromatography and to draw the breakthrough curve. The Br^- breakthrough curve has good reproducibility, indicating that the soil column system is stable and the hydrodynamic states of column experiments are similar (Fig. 2).

3.5 Breakthrough Curves of Bacteria

For the concentration of bacterial solution and plate counts, please *see* Subheading 3.3. For bioluminescence determination please *see* Subheadings 3.8, **steps 1–10** and 3.9, **step 2**. These instructions are not repeated here.

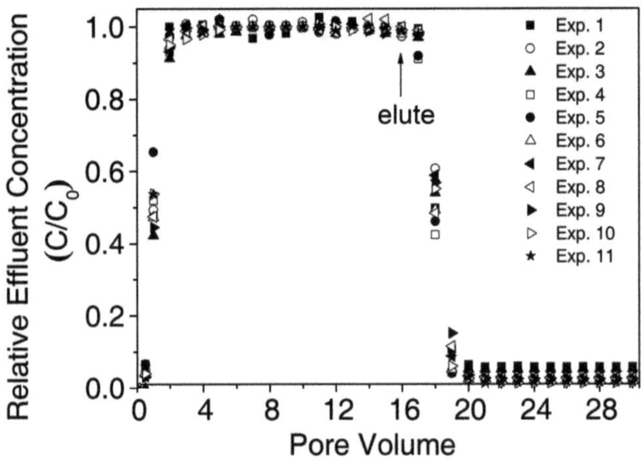

Fig. 2 Breakthrough curves of bromide

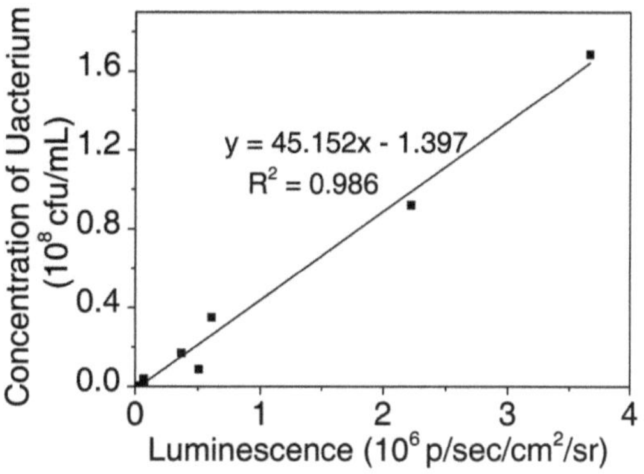

Fig. 3 Calibration curve between bioluminescence and plate counting of bacteria (*Pf*5RL) 2 h after the bacterial solution preparation

1. Prepare a standard curve based on the results of plate counting and bioluminescence determination. The original solution is diluted step by step, part of which is taken for plate counting, and the other part is put in the 96-well microtiter plate for bioluminescence determination. Obtain the standard curve with luminescence (p/s/cm^3 sr) as the *x*-coordinate and concentration of bacterium (cfu/mL) as the *y*-coordinate (Fig. 3).
2. After obtaining the standard curve, bioluminescence determination can be used to detect the high concentration of bacterial liquid.
3. Effluent samples (from Subheading 3.2, **steps 9** and **10**) are used to measure cell concentration by plate counting and bioluminescence determination in order to obtain the breakthrough curves of bacteria by plotting the relative effluent concentration (C/C_0) as a function of pore volumes of injected suspension or solution (Figs. 4, 5, and 6).

3.6 Retention Curve of Bacteria

The saturated porous media (from Subheading 3.2, **step 11**) are carefully excavated in about 1-cm thick increments and placed into 250-mL vials containing excess sterile background electrolyte solution. Then, the vials are slowly shaken for 15 min to liberate any reversibly retained bacteria. Finally, the bacterial concentrations in the excess solution are determined by plate counting [10]. The observed retention profiles for the bacteria are plotted by taking the percentage of residual bacteria at each depth relative to the total residual bacteria in the entire soil column as a function of travel distance from the column inlet (Fig. 7).

3.7 Haptotaxis of Bacteria

Chemotaxis and haptotaxis are important biological mechanisms that influence bacterial movement in soils. The IVIS bioluminescent imaging method can be used to assay the chemotaxis and haptotaxis of bacteria. Using naphthalene as a representative

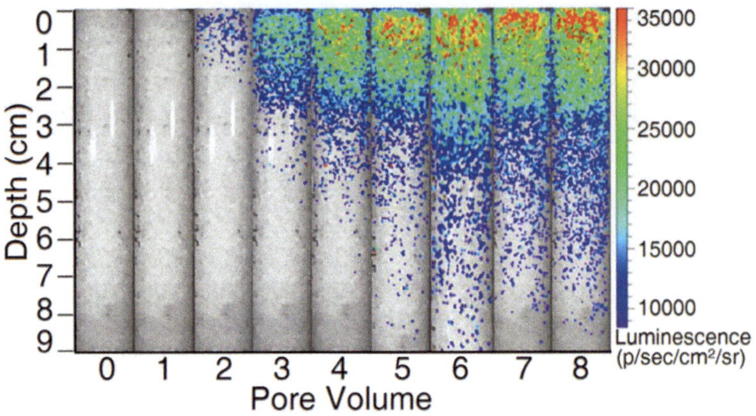

Fig. 4 Real-time distribution of bacteria (*Pf*5RL) at different pore volumes during bacterial transport downward through clean sand

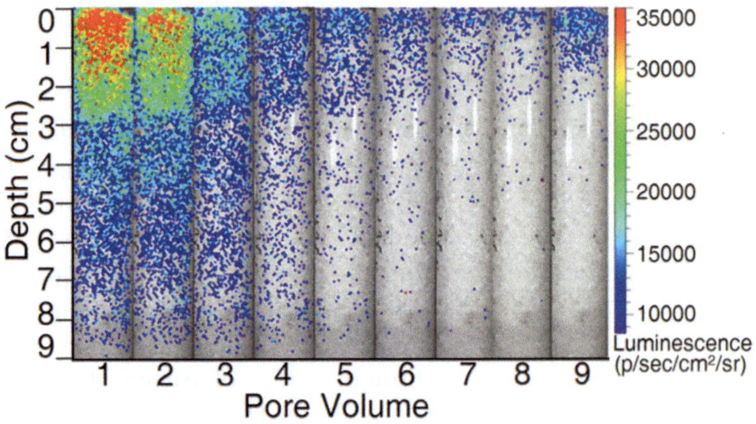

Fig. 5 Real-time distribution of bacteria (*Pf*5RL) at different pore volumes during downward elution from clean sand

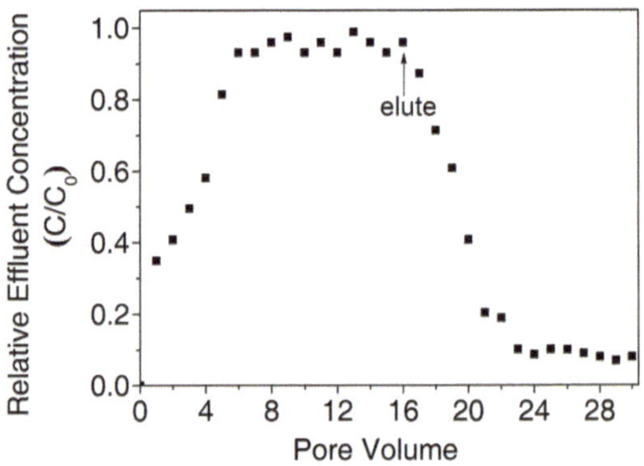

Fig. 6 Breakthrough curves of bacteria (*Pf*5RL)

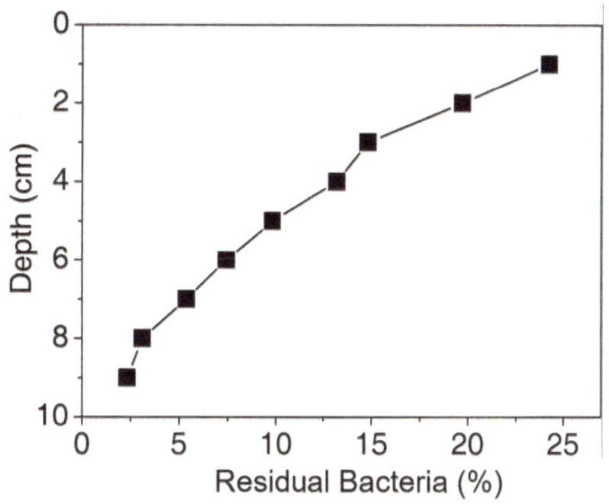

Fig. 7 Retention curve of bacteria (*Pf*5RL)

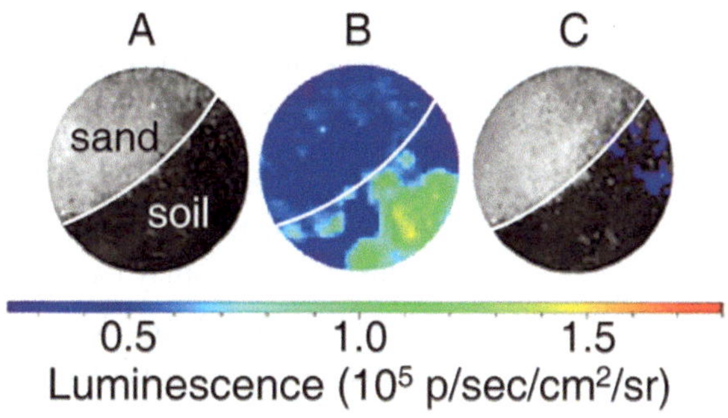

Fig. 8 Effect of surface-bound naphthalene (a representative chemoattractant) on the distribution of chemotactic bacteria (*Pf*5RL) in heterogeneous porous media during transport and elution processes. (**a**) Naphthalene-saturated column (0.5 cm in depth and 3.5 cm in diameter) packed half by half with a model sand ($d_{50} = 700$ μm) and a natural sandy soil ($d_{50} = 500$ μm). (**b**) Cotransport of naphthalene and bacteria through the column. (**c**) Retention of naphthalene and bacteria in soil (not in sand) after elution with background solution (10 mM NaCl)

chemoattractant, we compare the chemotactic and haptotactic movements of bacteria (*Pf*5RL) in a uniform, less reactive sand and a heterogeneous, reactive soil. Much more *Pf*5RL are adsorbed in the soil than in the sand during both transport and elution processes (Fig. 8). The elution removes all *Pf*5RL in the sand, but only a portion is removed from the soil. Continuous spreading of the retained *Pf*5RL in the soil after termination of the elution demonstrates that haptotaxis is a significant mechanism that reduces bacterial movement in the heterogeneous, reactive soils. Further detailed analysis can be carried out in relation to the breakthrough curves and profile distribution of bacteria in the soil columns.

3.8 Use of IVIS Spectrum Imaging System

1. Initialize the IVIS Spectrum.
2. Set temperature between 20 and 40 °C (*see* **Note 10**).
3. Place the column in the imaging chamber and close the door (*see* **Note 11**).
4. Start the Imaging Wizard and set up a Sequence.
5. Choose an imaging mode: Bioluminescence.
6. Choose an imaging option: Open Filter, Spectral Unmixing, DLIT (*see* **Note 12**).
7. Step through the rest of the wizard.
8. Put a check mark next to *Luminescent* and select *Auto* exposure (click the arrows) in the control panel. The software automatically determines the binning and F/Stop settings.
9. Put a check mark next to *Photograph* (optional).
10. Select a field of view (*see* **Note 13**).
11. Collect image in every pore volume (luminescence unit: photons/second/cm^2/steradian, with an abbreviation p/s/cm^2/sr).

3.9 Working with Images

1. Adjust Image Appearance: Use the Image Adjust tools to adjust image display (*see* **Note 14**).
2. View Intensity Data (Measuring Signal in 2D Image Data) at a particular (x, y) location along a line drawn on the image within a user-selected region of interest (ROI) (*see* **Note 15**).
3. Obtain information from the images.

4 Notes

1. Biologically engineer bacteria to produce stable bioluminescence. The luminous intensity of bacteria should meet the requirements of the detection system (IVIS Spectrum Imaging System or imaging system from other manufacturer).
2. Prior to use or coating, the sand is sequentially washed with HCl (10 mM) and then NaOH (10 mM) solutions to get rid of suspended impurities and finally rinsed with deionized water.
3. The sand is coated with humic acid (HA) using the following method [11]. First, 20 g of HA is dissolved in a plastic beaker with 200 mL of KOH (0.1 M) and stirred for 30 min. Next, 200 g of clean sand is added to the suspension and stirred vigorously for 10 min. The suspension is then allowed to

stand for approximately 8 h during which it is stirred for 10 min once every hour. Subsequently, 6 M HCl is titrated into the suspension under well-mixed conditions until the suspension pH reaches 2.0. The suspension is kept at 70 °C for 24 h and stirred periodically. Finally, the coated sand is washed using deionized water until there was no impurity in the supernatant.

4. The sand is coated with goethite using $FeCl_3$ by the following method [12]. 80 mL of a 2.5 M $FeCl_3$ solution is poured over 200 g clean sand, and the mixture is heated at 110 °C and stirred hourly for about 3 h, by which time it appears to be dry. The temperature is then raised to 550 °C for 3 h, after which the sand is cooled to room temperature in air. Finally, the coated sand is washed using deionized water until the black-colored fraction is washed away, remaining a dark red coating on the sand.
5. Columns made of acrylic sheets or other transparent materials must have no effect on microbial and bioluminescence imaging before use.
6. If the column cannot be sterilized by high temperature, thoroughly wipe with anhydrous alcohol.
7. Filling materials must be compact enough to avoid collapse after introducing liquid. The bulk density of sand and loam packings must be controlled above 1.7 g/cm^3 and 1.1 g/cm^3 or close to the natural field conditions.
8. The column experiments are performed at 25 °C in replicates under steady flow conditions at a given pore velocity.
9. Avoid liquid exchange between different soil depths during sampling.
10. The stage is temperature-controlled to keep subjects warm during imaging. The default temperature is 37 °C and is self-monitoring after the system is initialized. The imaging stage may be set to a temperature from 20 to 40 °C.
11. The maximum size of the image cannot exceed 20 cm, otherwise it affects the function of the lens. The connecting teflon tubes must be very thin to ensure that the chamber door is closed well for high-quality images. Advance tests on the effect of door closing are needed.
12. Imaging Wizard—Imaging Mode Options for IVIS Spectrum.

Imaging mode	Options
Bioluminescence	*Open Filter*—Acquire a luminescent image at maximum sensitivity *Spectral Unmixing*—Acquire an image sequence for analysis using the Spectral Unmixing tools which use a mathematical operation to separate the signals from multiple luminescent probes *DyCE*—Acquire a time series of optical images following a bolus injection of probe (radiotracer, bioluminescent, or fluorescent) to track probe biodistribution *DLIT* (diffuse light tomography)—Acquire an image sequence for analysis with the DLIT algorithm that reconstructs the position, geometry, and strength of 3D luminescent sources

13. Field of View (FOV) Settings—IVIS Spectra are provided below.

FOV setting	FOV (cm)
A	4
B	6.5
C	13
D	22.5

14. Most of the Image Adjust tools do not change the image data (for example, adjusting the color scale or color table). However, binning and smoothing may slightly change image data, and therefore should only be applied after image data have been analyzed.
15. Please refer to the values in the "average" list when you use ROI, which we believe is more accurate.

Acknowledgments

This work is partly supported by the National Natural Science Foundation of China (Grant No. 41730858).

References

1. Quintelas C, Ferreira EC, Lopes JA, Sousa C (2018) An overview of the evolution of infrared spectroscopy applied to bacterial typing. Biotechnol J 13(1):170044910. https://doi.org/10.1002/biot.201700449
2. Jia M, Li SM, Zang LG, Lu XN, Zhang HY (2018) Analysis of biomolecules based on the surface enhanced Raman spectroscopy. Nanomaterials (Basel) 8(9):E730. https://doi.org/10.3390/nano8090730
3. Yang X, Zhao X, Yang K, Liu YP, Liu Y, Fu WL, Luo Y (2016) Biomedical applications of terahertz spectroscopy and imaging. Trends

Biotechnol 34(10):810–824. https://doi.org/10.1016/j.tibtech.2016.04.008

4. Yang X, Yang K, Luo Y, Fu WL (2016) Terahertz spectroscopy for bacterial detection: opportunities and challenges. Appl Microbiol Biotechnol 100(12):5289–5299. https://doi.org/10.1007/s00253-016-7569-6
5. Sivakumar P, Fernandez-Bravo A, Taleh L, Biddle JF, Melikechi N (2015) Detection and classification of live and dead *Escherichia coli* by laser-induced breakdown spectroscopy. Astrobiology 15(2):144–153. https://doi.org/10.1089/ast.2014.1181
6. Saari S, Jarvinen S, Reponen T, Mensah-Attipoe J, Pasanen P, Toivonen J, Keskinen J (2016) Identification of single microbial particles using electro-dynamic balance assisted laser-induced breakdown and fluorescence spectroscopy. Aerosol Sci Technol 50 (2):126–132. https://doi.org/10.1080/02786826.2015.1134764
7. Elmasry G, Kamruzzaman M, Sun DW, Allen P (2012) Principles and applications of hyperspectral imaging in quality evaluation of agro-food products: a review. Crit Rev Food Sci Nutr 52(11):999–1023. https://doi.org/10.1080/10408398.2010.543495
8. He HJ, Sun DW (2015) Hyperspectral imaging technology for rapid detection of various microbial contaminants in agricultural and food products. Trends Food Sci Technol 46 (1):99–109. https://doi.org/10.1016/j.tifs.2015.08.001
9. Wang KQ, Pu HB, Sun DW (2018) Emerging spectroscopic and spectral imaging techniques for the rapid detection of microorganisms: an overview. Compr Rev Food Sci Food Saf 17 (2):256–273. https://doi.org/10.1111/1541-4337.12323
10. Bai HJ, Cochet N, Pauss A, Lamy E (2016) Bacteria cell properties and grain size impact on bacteria transport and deposition in porous media. Colloids Surf B Biointerfaces 139:148–155. https://doi.org/10.1016/j.colsurfb.2015.12.016
11. Zhuang J, Jin Y (2003) Virus retention and transport as influenced by different forms of soil organic matter. J Environ Qual 32 (3):816–823
12. Benjamin MM, Sletten RS, Bailey RP, Bennett T (1996) Sorption and filtration of metals using iron-oxide-coated sand. Water Res 30 (11):2609–2620. https://doi.org/10.1016/S0043-1354(96)00161-3

Part II

In Vivo Imaging of Microbial Pathogenesis

Chapter 6

In Vivo Bioluminescent Imaging of *Yersinia ruckeri* Pathogenesis in Fish

José A. Guijarro and Jessica Méndez

Abstract

Bioluminescent reporters and advanced luciferase technologies are useful to study host–pathogen interactions. This chapter describes the use of the *luxCDABE* operon from *Photorhabdus luminescens* as a tool to analyze the progression of the fish pathogen *Yersinia ruckeri* during the infection of rainbow trout, as well as the quantification of promoter activity of specific bacterial genes during host colonization.

Key words *Yersinia ruckeri*, Rainbow trout, Infection process, Bioluminescence imaging, *luxCDABE* operon, IVIS® Imaging System, Promoter activity

1 Introduction

Biological systems producing different kinds of light have been used for the construction of different tools useful for studies such as the identification of promoters, the study of gene expression or the detection of bacteria in environmental samples and food [1]. One of the main applications of these systems has been the analysis of the bacteria–host interaction during the development of an infection process, such as *Salmonella* Typhimurium infection in mice, which was the first interaction analyzed by using the *luxCDABE* operon [2]. Since then, the system has been also used in gram-positive bacteria such as *Staphylococcus aureus* [3] and *Streptococcus pneumoniae* [4]. For these analyses, the luciferase-based systems, specifically those using the *luxCDABE* operon from *Photorhabdus luminescens* [5], have advantages when compared, for example, with GFP or similar substantiated tools, because they do not need previous excitation for light emission, host tissue autofluorescence noise is avoided and luciferase average life is shorter, which makes it more appropriate for real-time tracking.

Yersinia ruckeri is the causative agent of the "enteric redmouth disease" (ERM), which mainly affects salmonids [6, 7]. We have

Steven Ripp (ed.), *Bioluminescent Imaging: Methods and Protocols*, Methods in Molecular Biology, vol. 2081,
https://doi.org/10.1007/978-1-4939-9940-8_6,

applied these techniques to follow the bacterial progression into the fish by engineering strains that carry the *luxCDABE* operon from *Photorhabdus luminescens* [5]. In this chapter, we describe how *Y. ruckeri-lux* strains were used to establish infections in rainbow trout and how the bacterial dissemination was monitored using an IVIS® Imaging System (PerkinElmer) [8]. In addition, promoter probe vector based on the *luxCDABE* operon allowed for the study of bacterial gene expression inside the fish.

2 Materials

2.1 Media, Molecular Reagents and Solutions

1. Tryptone soy broth (TSB): 17.0 g/L tryptone, 3.0 g/L soytone, 2.5 g/L glucose, 5.0 g/L NaCl, 2.5 g/L K_2HPO_4.
2. Tryptone soy agar (TSA): TSB with 1.5% agar added.
3. M9C: minimal medium M9 [9] supplemented with 0.2% casamino acids and 1.5% agar.
4. Phosphate buffered saline (PBS): 8 g/L NaCl, 1.44 g/L Na_2HPO_4, 0.24 g/L K_2HPO_4, 0.2 g/L KCl prepared in distilled water and sterilized by autoclaving.
5. TAE buffer: 40 mM Tris acetate, 1 mM EDTA, pH 8.3.
6. Molecular reagents: Restriction enzymes (*Bam*HI, *Xho*I, and *Eco*RI), T4 DNA ligase, High-fidelity Taq polymerase kit.
7. DNA primers: primers cdsABTER-F (5′-ATGCCTCGAGACTTAATGGTATTCAGA-3′) and cdsABTER-R (5′-ATGCGGATCCATGCATGCGAATTCTTCTACCAGTGCAACGG-3′) for targeting the *Y. ruckeri cdsAB* operon promoter (*Xho*I, *Bam*HI, and *Eco*RI restriction sites are underlined).
8. Kanamycin: 50 mg/mL stock concentration of kanamycin filtered through a 0.2 μm nitrocellulose filter.
9. Glycerol: 10% solution prepared in distilled water and sterilized by autoclaving.

2.2 Bacterial Strains and Vectors

1. Bacterial strains: *Y. ruckeri* 150 (J.L. Larsen, Denmark) and *Escherichia coli S17-1λpir* [10].
2. Vector: pCS26-Pac [11].

2.3 Other Materials

1. Experimental fish: fish (rainbow trout or other desired species) can be obtained from a fish farm. It is very important that they are in good health conditions and free of any antibiotic treatment, vaccination and without any ERM episodes. They should have a weight between 6 and 8 g.
2. IVIS® Imaging System (PerkinElmer).
3. 1 mL syringes and syringe needles.

4. Petri dishes, sterile (100 mm × 15 mm).
5. 96-well microtiter plates with white walls.

3 Methods

3.1 Preparation of Electrocompetent Y. ruckeri *and* Escherichia coli *S17-1λpir Cells*

1. Subculture 2 mL of a fresh overnight culture into 100 mL TSB in a 250 mL flask. Grow cells on a 28 °C (for *Y. ruckeri*) or 37 °C (for *E. coli*) shaker to midlog phase (optical density at 600 nm (OD_{600}) = 0.5–0.7). Then chill cells on ice for 20 min. For all subsequent steps, keep the cells close to 0 °C and chill all containers in ice before adding the cells.
2. Transfer the culture to 50 mL polypropylene centrifuge tubes and centrifuge the cells at 5432 × *g* for 15 min at 4 °C.
3. Discard the supernatant carefully and resuspend the cells in the same volume of cold sterile milliQ ultrapure distilled water. Then, centrifuge the cells at 5432 × *g* for 15 min at 4 °C.
4. Discard the supernatant carefully and resuspend the cells in 1/2 volume of cold sterile milliQ water. Then, centrifuge the cells at 5432 × *g* for 15 min at 4 °C.
5. Discard the supernatant carefully and resuspend the cells in 1 mL cold sterile 10% glycerol and centrifuge the cells at 5432 × *g* for 15 min at 4 °C.
6. Discard the supernatant and resuspend the cells in 0.1 mL cold sterile 10% glycerol and prepare aliquots of 40 μL on dry ice and store at −80 °C.

3.2 Introduction of pCS26-Pac and Derivatives into Y. ruckeri *or* E. coli *S17-1λpir Cells*

1. Add 1–3 μL of pCS26-Pac DNA or its derivatives (approximately 60 ng) to 40 μL of electrocompetent cells of *Y. ruckeri* or *E. coli S17-1λpir*, previously thawed on ice. Mix gently with a pipette, incubate 1 min on ice and transfer the sample to a chilled 0.2 mm electroporation cuvette. Tap the liquid to the bottom of the cuvette (*see* **Note 1**).
2. Give an electrical pulse to generate pores in the cell membranes and allow the entry of DNA. Electroporation should be carried out at a voltage of 2.5 kV and the time constant should be about 5.0 ms (*see* **Note 2**).
3. After electroporation, remove quickly the cuvette from the electroporation chamber and add 1 mL TSB to the cuvette. Mix gently and quickly with a Pasteur pipette and transfer the cell suspension to a 10 mL polypropylene tube. This step should be done quickly after shocking the cells because cell viability will decrease over time. Incubate cells for 120 min on a 28 °C shaker (for *Y. ruckeri*) or for 60 min on a 37 °C shaker

(for *E. coli S17-1λpir*) to allow for cell recovery and expression of the antibiotic resistance gene.

4. Then, plate the sample (use different volumes, for example 0.1 and 0.9 mL) on a selective medium (TSA media supplemented with 50 μg/mL of kanamycin) and incubate for 24 h at 28 °C (for *Y. ruckeri*) or 37 °C (for *E. coli*).
5. Verify the introduction of the pCS26-PAC vector in *Y. ruckeri* or *E. coli S17-1λpir* cells by plasmid DNA isolation using a commercial kit and standard molecular methodologies such as restriction digestion (*see* **Note 3**).

3.3 Preparation of Rainbow Trout for Infection

1. Place fish and water from the fish farm in portable tanks and transport them with appropriate aeration to the laboratory.
2. After transport, allow fish to acclimatize for at least 24 h to the laboratory temperature, which has to be previously fixed at 18 °C (*see* **Note 4**). Then, distribute fish in groups of 30–40 individuals into several 250 L tanks containing 140 L of dechlorinated water, where they are maintained during at least 2 days under starvation conditions (prior to any experiment).

3.4 Challenge of Rainbow Trout by Intraperitoneal Injection of Y. ruckeri Strains

1. Grow the *Y. ruckeri* strain of interest up to OD_{600} of 0.5 (approximately 10^8 colony forming units (cfu)/mL) in TSB, in a 250 rpm shaker and at 18 °C, to match the temperature at which ERM outbreaks occur, and prepare tenfold serial dilutions in PBS.
2. Inoculate by intraperitoneal injection of 0.1 mL of dilution 10^{-2} (containing 10^6 cfu/mL) into thirty rainbow trout weighing from 6 to 8 g and previously anesthetized by immersing them on ice water (hypothermic anesthesia).
3. Check that the injected fish recovers from hypothermia and injection after 60 s and that they are swimming normally. Any fish that does not recover within 2–3 min or is swimming erratically should be removed and euthanized.
4. Observe the fish for 6 days, with daily removal of dead individuals.

3.5 Immersion Challenge of Rainbow Trout with Y. ruckeri Strains

1. Grow *Y. ruckeri* strain of interest up to OD_{600} of 0.5 (approximately 10^8 cfu/mL) at 18 °C in TSB with 250 rpm.
2. Subdivide 1 L of the culture into 250 mL centrifuge bottles and centrifuge the cells at 4629 × *g* for 15 min at 4 °C. Discard carefully the supernatant and wash the cells once with sterile distilled water.
3. Resuspend the four pellets (corresponding to the four bottles) in 250 mL of sterile distilled water and transfer the suspension

to a tank containing 9.75 L of dechlorinated water at 18 °C. The concentration of cells in the tank should be about 10^7 cfu/mL.

4. Introduce thirty fish into this tank for an hour to expose them to *Y. ruckeri* strain of interest. Then, remove the fish and transfer them to a 250 L tank containing 140 L of dechlorinated water at 18 °C.
5. Observe the fish for 6 days, with daily removal of dead individuals.

3.6 Visualization of Y. ruckeri Infection in Rainbow Trout

The vector pCS26-Pac [11] can be used directly, without additional modifications, to label pathogenic bacteria and monitor bacterial dissemination throughout the course of infection in rainbow trout.

1. Introduce pCS26-Pac plasmid in *Y. ruckeri* by electroporation (*see* Subheading 3.2) to obtain *Y. ruckeri* pCS26-Pac.
2. Infect thirty fish by intraperitoneal injection (*see* Subheading 3.4) and thirty fish by immersion (*see* Subheading 3.5) with *Y. ruckeri* pCS26-Pac.
3. Take five surviving fish from each challenge (intraperitoneal injection and immersion) at 24 h intervals over a 3 day period and remove the dead fish every day.
4. Sacrifice by euthanasia the surviving fish with an anesthesia overdose. To do so, place fish into a bath of ethylene glycol monophenyl ether solution at a concentration of 0.5–0.6 mL/L.
5. Dissect euthanized fish, after removal from anesthesia, and fish dead from infection.
6. Analyze the fish with an IVIS® Imaging System (PerkinElmer) to monitor bacterial dissemination inside the fish.
7. Start the IVIS Lumina, initialize the Living Image software and allow 10–15 min for the camera to reach the proper temperature, indicated by the change of light color (from red to green) of the corresponding indicator.
8. Place each dissected fish in a sterile petri dish and introduce it into the IVIS's imaging chamber, at the center of the stage, and select the exposure time and binning. If unsure of what exposure time to use, it is best to start with low sensitivity settings (10 s, binning medium) and increase as necessary.
9. Click the Acquire button on the control panel to get IVIS image. After the exposure is completed the overlaid image is displayed. Confirm that the signal of interest is above 600 counts and below 60,000 counts. If the signal level is unacceptable, adjust Exposure Time or Binning and repeat the image acquisition. Save all the acquired images in a folder for further analysis.

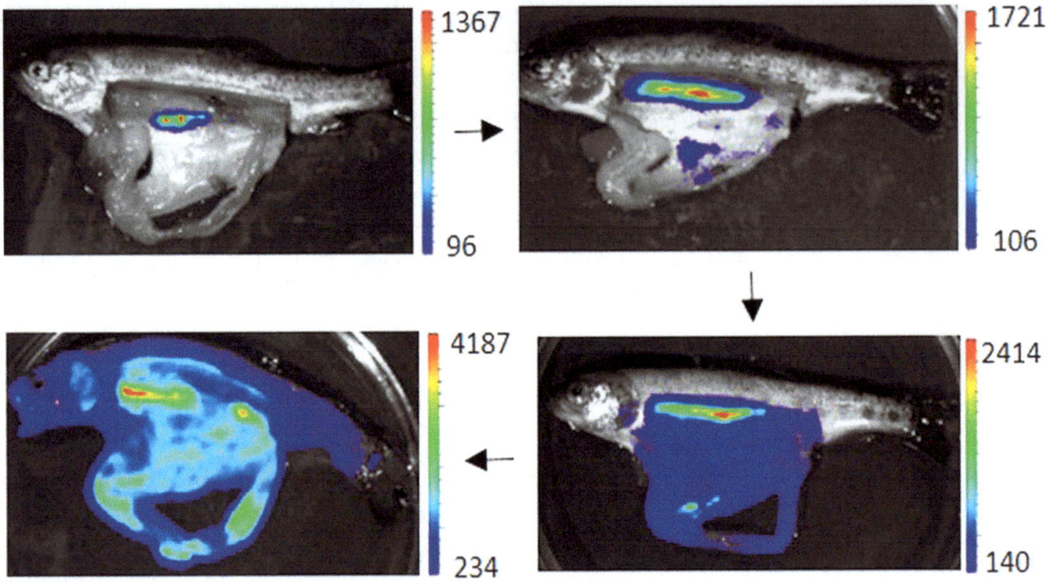

Fig. 1 Bioluminescent tracking of *Y. ruckeri* pCS26-Pac infection in rainbow trout after intraperitoneal injection with 10^4 cfu per fish. Color bars on the right represent scale for light quantification

10. Adjust the scale bar for each image. Max Bar and Min Bar set the maximum and minimum values, respectively, for the range of data associated with the color bar for the image. The minimum value serves as a threshold below which data in luminescent images are not displayed.
11. Compile the images obtained and the time postinfection they were taken at to make a composition of the bacterial dissemination during the infection process (Fig. 1) (*see* **Note 5**).
12. To analyze individual organs, draw the region of interest (ROI) over the organ of interest, then click "Measure organs" under ROI Tools in the Tool Palette. Total relative luminescent units (RLU) will be shown on the image as well as in a separate table (Fig. 2). These values can be exported directly to Microsoft Excel or other database software for further studies.

3.7 Transcriptional Terminator Cloning into pCS26PAC to Generate a Reporter System

Since the pCS26-Pac [11] plasmid itself yields high levels of bioluminescence, it cannot be used directly to quantify the expression of *Y. ruckeri* genes. For this reason, it is necessary to introduce a strong transcriptional terminator to avoid read-through from promoters upstream of the *luxCDABE* operon [8].

1. Target the promoter from the *cdsAB* operon of *Y. ruckeri* [8] for PCR amplification using the primers cdsABTER-F (5′-ATGCCTCGAGACTTAATGGTATTCAGA-3′) and cdsA BTER-R (5′-ATGCGGATCCATGCATGCGAATTCTTCTA CCAGTGCAA CGG-3′) (Fig. 3a). The former primer contains

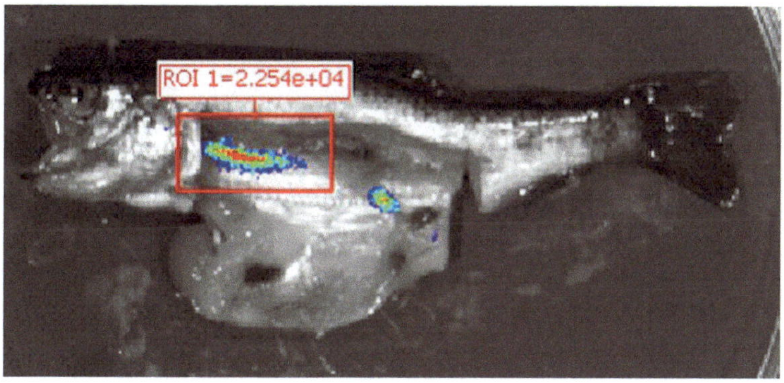

Fig. 2 Example of a region of interest (ROI) drawn over a fish organ of interest. Relative luminescent units (RLU) are shown on the image

a *Xho*I restriction site and the latter contains *Bam*HI and *Eco*RI restriction sites (underlined sequences). Genomic DNA from *Y. ruckeri* was used as template.

2. Check the length of the amplicon product on a 1% agarose gel. Then, using a clean scalpel, ultraviolet light and minimal exposure time, cut out as small an agarose band as possible containing the sample of interest. Purify the DNA from the agarose band with a commercial kit following the manufacturer's instructions.
3. Digest the resulting PCR fragment containing the transcriptional terminator with *Xho*I and *Bam*HI restriction enzymes and ligate it (using a T4 DNA ligase) to the vector pCS26-Pac previously digested with the same enzymes and purified (Fig. 3b).
4. Introduce the ligation mixture in *E. coli* S17-1λ*pir* by electroporation (*see* Subheading 3.2) and select the transformants on TSA supplemented with 50 μg/mL of kanamycin. The IVIS Lumina equipment can also be used to search for nonluminescent cells (*see* **Note 6**).
5. Select a clone of *E. coli* harboring the pCS26-Pac with the transcriptional terminator (pCS26-PacTER) and obtain the plasmid DNA using commercial kits.
6. Introduce the pCS26-PacTER vector in competent cells of *Y. ruckeri* by electroporation (*see* Subheading 3.2).
7. Use the IVIS Lumina equipment to select the appropriate clones of *Y. ruckeri* carrying pCS26-Pac-TER, which are those lacking bioluminescence (Fig. 3c).

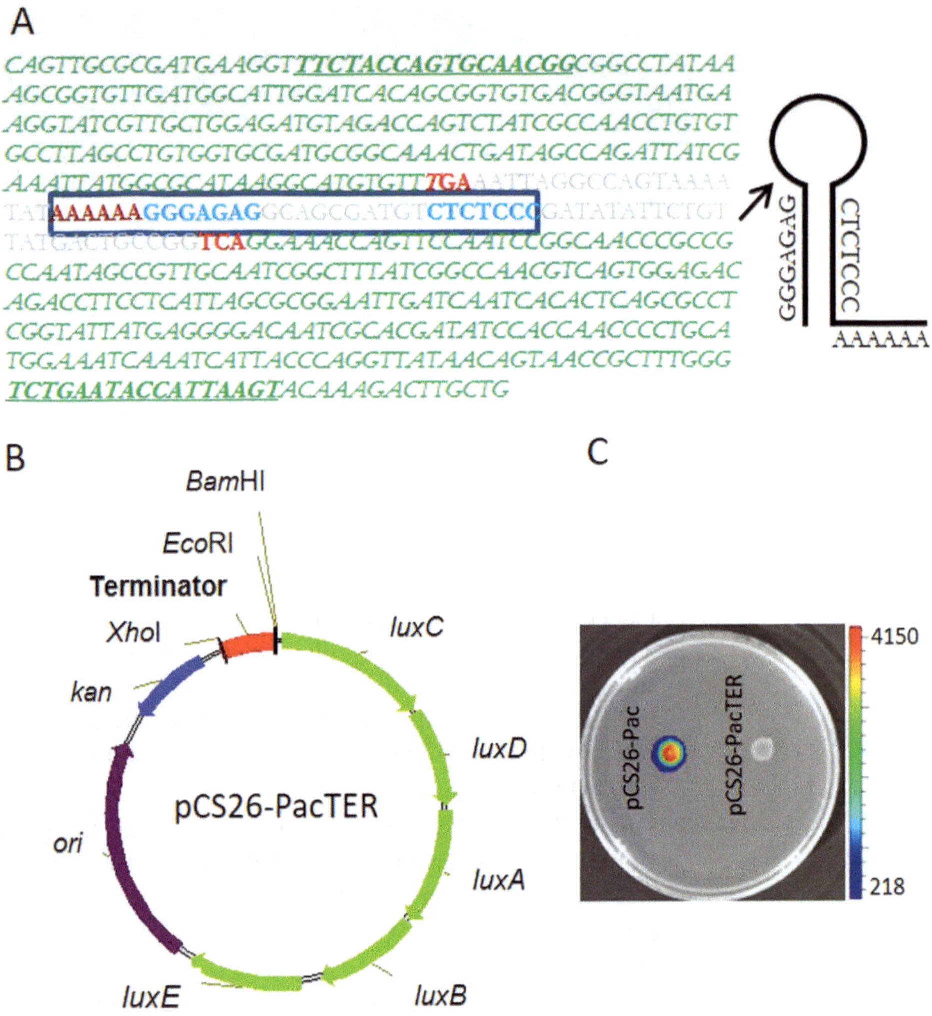

Fig. 3 Construction of the pCS26Pac-TER plasmid. (**a**) Sequence of the transcriptional terminator of the *cdsAB* operon. Primers used to amplify this region are underlined. (**b**) Plasmid pCS26-PacTER (**c**) Bioluminescence imaging of *Y. ruckeri* harboring pCS26-Pac and pCS26-PacTER plasmids. Color bars on the right represent the scales for light quantification

3.8 Promoter Amplification and Cloning into pCS26Pac-TER

1. Identify the promoter sequence upstream of the ATG of the gene of interest and design the appropriate primers to amplify that sequence by PCR. Add an *Eco*RI restriction site to the 5′ end of the forward primer and a *Bam*HI restriction site to the 5′ end of the reverse primer. An extension of the primer with four additional random nucleotides before the restriction site facilitates restriction enzyme binding and cleavage.
2. Check the length of the amplicon product on a 1% agarose gel and proceed as in Subheading 3.7, **step 2**.

3. Digest the PCR fragment with the *Bam*HI and *Eco*RI enzymes and ligate it to pCS26PAC-TER previously digested with the same enzymes by using a T4 DNA ligase according to the manufacturer's instructions.
4. Introduce the ligation mixture into *E. coli* S17-1λ*pir* by electroporation (*see* Subheading 3.2) and select the transformants on TSA supplemented with 50 μg/mL of kanamycin.
5. Verify the introduction of the construction of interest into *E. coli S17-1λpir* cells by plasmid DNA isolation using a commercial kit and standard molecular methodologies as restriction digestion, PCR or sequencing.
6. Introduce the construction of interest in *Y. ruckeri* by electroporation (*see* Subheading 3.2) and subsequent selection on TSA supplemented with 50 μg/mL of kanamycin. Verify the result of the transformation experiment as described in previous steps.

3.9 In Vitro Detection of Promoter Activity of Y. ruckeri Genes

Gene expression can be quantified through the emission of luminescence of *Y. ruckeri* strains both in solid and liquid media.

1. For expression on solid medium, spot 1 or 2 μL of overnight cultures of the strain harboring the plasmid pCS26PAC-TER with the promoter sequence of interest (e.g., the promoter of the gene encoding the Yrp1 protease of *Y. ruckeri;* strain called *Y. ruckeri* Pyrp1) and *Y. ruckeri* pCS26PAC-TER (used as negative control) under the appropriate conditions (e.g., on M9C medium at 18 °C) (*see* **Note** 7) and image the colonies with the IVIS Imagen system (Fig. 4). Several promoters can be studied at the same time, but it is recommended to use a separate plate with the corresponding negative control for each gene. If the same plate is used an ROI for each colony should be drawn.
2. For expression in liquid media, use 96-well microtiter plates with white walls containing the appropriate medium as well as the relevant controls and replicas. Dilute 1:100 stationary-phase cultures of each strain in the appropriate fresh medium at a specific temperature (consider the optimal induction conditions for each promoter). Grow cells to a specific OD_{600} (0.4–0.6) and prepare serial dilutions. Transfer 0.1 mL aliquots, at least by triplicate, to the wells of the microtiter plate. *Y. ruckeri* pCS26-PacTER should be used as negative control. A constitutive promoter such as the one from *gyrB* gene can be used as reference [8]. Several promoters can be studied at the same time. In this case, Grid ROIs are particularly useful for microtiter well plate measurements. When Grid is selected in the ROI shapes menu a pull-down listing appears that shows standard well plate sizes (e.g., 12 × 8 and 24 × 16).

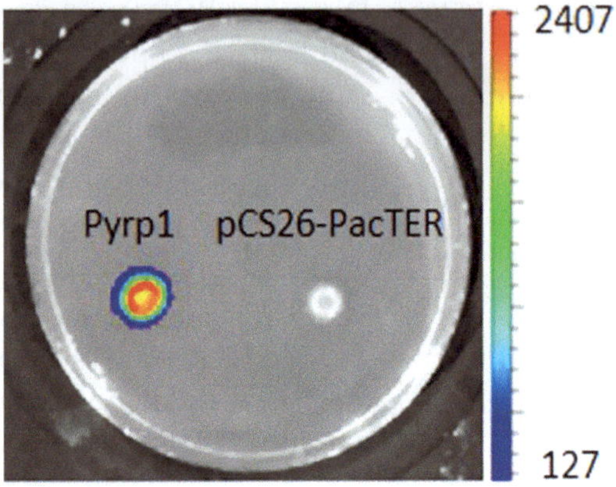

Fig. 4 Bioluminescence emitted by *Y. ruckeri* Pyrp1 after 24 h of incubation on M9C medium at 18 °C. On the right, the colony of *Y. ruckeri* pCS26-PacTER, used as negative control of luminescence, is shown

3.10 In Vivo Detection of Promoter Activity of Y. ruckeri Genes

1. Grow overnight cultures of *Y. ruckeri* strains of interest, including *Y. ruckeri* pCS26PAC-TER as negative control and *Y. ruckeri* pCS26-Pac with the promoter of a housekeeping gene (e.g., *gyrB*) in TSB up to an OD_{600} of 0.5 and prepare serial dilutions in PBS.
2. For each strain, introduce 0.1 mL of dilution 10^{-2} (10^5 cfu per fish) into groups of 25 fish by intraperitoneal injection (*see* Subheading 3.4 and **Note 8**).
3. Collect ten dead fish/strain at 48 h postinfection.
4. Extract the swim bladder, or any other organ of reference, from each fish and analyze with the IVIS Lumina system (3 min of exposition and binning 2). Annotate total RLU for each organ. Several swim bladders can be simultaneously imaged using the ROI tool.
5. After the images have been taken, place each swim bladder in 200 μL of PBS and shake vigorously. Take a 0.1 mL aliquot of the suspension to prepare tenfold dilutions in PBS.
6. Use 0.1 mL from dilution 10^{-5} to inoculate plates of TSA containing 50 μg/mL of kanamycin.
7. After 24 h of incubation at 28 °C, count the resulting colonies to calculate the number of bacteria per swim bladder. Finally, estimate RLU per 1000 cfu.

4 Notes

1. It is very important to maintain the competent cells on ice all the time until electroporation.
2. Plasmid purity has an effect on transformation efficiency. In general, this is high when plasmids are purified with commercial kits.

3. The IVIS Lumina equipment can be used for a rapid screening of colonies of interest (luminescent or nonluminescent cells).
4. The time necessary for the fish to acclimate to the laboratory temperature will depend on the water temperature of the fish farm. If this is very cold, acclimation should be done by increasing slowly the water temperature at 2° intervals until reaching 18 °C.
5. Considering the detection limit of the IVIS Imaging System [8], the number of bacteria during the first 12 h postinfection may be below the detection level and sites of entry in which bacterial replication is limited could appear as nonluminescent.
6. *E. coli* is used as a host to clone heterologous genes because of its easy transformation of the exogenous DNA in relation to *Y. ruckeri*.
7. For induction of each specific promoter it is necessary to consider the optimal conditions for expression (medium, temperature, oxygen availability, etc.).
8. Bath infection procedure is considered as a more natural route of bacterial administration but the intraperitoneal injection has the advantage of enabling the standardization of the bacterial dose administered to each fish. For this reason, in vivo studies that involve a high number of individuals or require statistical analysis are usually carried out with this kind of infection.

References

1. Waidmann MS, Bleichrodt FS, Laslo T, Riedel CU (2011) Bacterial luciferase reporters: the Swiss army knife of molecular biology. Bioeng Bugs 2:8–16. https://doi.org/10.4161/bbug.2.1.13566
2. Contag CH, Contag PR, Mullins JI, Spilman SD, Stevenson DK, Benaron DA (1995) Photonic detection of bacterial pathogen in living host. Mol Microbiol 18:593–603. https://doi.org/10.1111/j.1365-2958.1995.mmi18040593.x
3. Francis KP, Joh D, Bellinger-Kawahara C, Hawkinson MJ, Purchio TF, Contag PR (2000) Monitoring bioluminescent *Staphylococcus aureus* infections in living mice using a novel *luxABCDE* construct. Infect Immun 68:3594–35600
4. Francis KP, Yu J, Bellinger-Kawahara C, Joh D, Hawkinson MJ et al (2001) Visualizing pneumococcal infections in the lungs of live mice using bioluminescence *Streptococcus pneumoniae* transformed with a novel Gram-positive *lux* transposon. Infect Immun 69:3350–3358. https://doi.org/10.1128/IAI.69.5.3350-3358.2001
5. Meighen EA, Szittner RB (1992) Multiple repetitive elements and organization of the *lux* operons of luminescent terrestrial bacteria. J Bacteriol 174:5371–5381. https://doi.org/10.1128/jb.174.16.5371-5381.1992
6. Fernandez L, Mendez J, Guijarro JA (2007) Molecular virulence mechanisms of the fish pathogen *Yersinia ruckeri*. Vet Microbiol 125:1–10. https://doi.org/10.1016/j.vetmic.2007.06.013
7. Guijarro JA, Garcia-Torrico AI, Cascales D, Mendez J (2018) The infection process of *Yersinia ruckeri*: reviewing the pieces of the jigsaw puzzle. Front Cell Infect Microbiol 8:218. https://doi.org/10.3389/fcimb.2018.00218
8. Mendez J, Guijarro JA (2012) *In vivo* monitoring of *Yersinia ruckeri* in fish tissues: progression and virulence gene expression. Environ Microbiol Rep 5:179–185. https://doi.org/10.1111/1758-2229.12030
9. Romalde JL, Conchas RF, Toranzo AE (1991) Evidence that *Yersinia ruckeri* possesses a high

affinity iron uptake system. FEMS Microbiol Lett 80:121–126. https://doi.org/10.1111/j.1574-6968.1991.tb04647.x

10. Simon R, Priefer U, Pühler A (1983) A broad host range mobilization system for *in vivo* genetic engineering: transposon mutagenesis in gram negative bacteria. Nat Biotechnol 1:784–791. https://doi.org/10.1038/nbt1183-784

11. Bjarnason J, Southward CM, Surette MG (2003) Genomic profiling of iron-responsive genes in *Salmonella enterica* serovar Typhimurium by high-throughput screening of a random promoter library. J Bacteriol 185:4973–4982. https://doi.org/10.1128/JB.185.16.4973-4982.2003

Chapter 7

In Vivo Bioluminescent Monitoring of Parasites in BALB/c Mouse Models of Cutaneous Leishmaniasis Drug Discovery

Diana Caridha, Susan Leed, and Alicia Cawlfield

Abstract

Confirming the in vivo efficacy of potential antileishmanial compounds that display in vitro potency and good chemical characteristics is one of the most important steps in preclinical research drug discovery before human clinical trials begin. Here we describe the use of the in vivo bioluminescent monitoring of high and low inocula of luciferase-expressing *Leishmania major* (*L. major*) parasites in traditional and more innovative rodent models of in vivo cutaneous leishmaniasis (CL) drug discovery.

Key words Cutaneous leishmaniasis (CL), Rodent models, In vivo imaging, IVIS, Bioluminescence signal, Luciferase-expressing *Leishmania major*, Drug efficacy, Base of the tail, Footpad, Ear infections

1 Introduction

The World Health Organization (WHO) estimates an incidence of 2 million new cases of leishmaniasis per year (0.5 million cases of visceral leishmaniasis and 1.5 million cases of CL), with more than 350 million people being at risk of infection [1]. There are currently no effective vaccines for this disease and the need to develop effective, safe, and fast-acting therapeutics is greater than ever [2, 3].

Currently, the vast majority of leishmaniasis in vivo models use a needle challenge of parasites alone and in large quantities. Even though they do not fully represent the CL pathology in humans, the in vivo lesion cure (dorsal and footpad) models using the BALB/c mouse and Golden Syrian Hamster (GSH) with *Leishmania* spp. as a source of CL infection have been widely used to test the efficacy of antileishmanial drugs [2, 4–8]. However, *Leishmania* is transmitted to animals and humans by various species of sand fly vectors. In a natural sand fly infection, 100 to 100,000 metacyclic *Leishmania* parasites together with approximately 40 salivary proteins may be injected intradermally (ID) [9, 10]. When injected along with *Leishmania*, the presence of extracted sand fly

Steven Ripp (ed.), *Bioluminescent Imaging: Methods and Protocols*, Methods in Molecular Biology, vol. 2081,
https://doi.org/10.1007/978-1-4939-9940-8_7,

salivary glands has shown to exacerbate infection and duration of the ulcer compared to *Leishmania* alone or no difference, warranting further research into the area of sand fly saliva–parasite–host [11, 12].

CL in vivo models were vastly improved by the integration of reporter genes into the *Leishmania* species genome to monitor intracellular proliferation of the parasites in longitudinal studies without the need to euthanize animals. The firefly luciferase gene (LUC) is a reporter gene that has been stably integrated within the parasite genome and has been used to monitor the in vivo intracellular proliferation of *Leishmania* spp. [13–15]. Multiple sources have reported a direct correlation between parasite numbers and luciferase activity [13, 15–17]. In a recent study [18], it was also determined that use of bioluminescent parasites to quantify *Leishmania* spp. infection in the ear is a more sophisticated and accurate approach compared with the more traditional measurements of lesion diameter, volume, and thickness. This is an ideal refinement for conducting longitudinal studies in a more noninvasive in vivo drug screening model, and parasite–host interaction studies [16]. New antileishmanial therapies for CL have been developed and tested using *Leishmania* spp. infections in the footpad or dorsal tail of the BALB/c mouse or GSH [4, 17, 19, 20]. Here we describe the use of in vivo bioluminescent monitoring parasites in models of CL drug discovery conducted in BALB/c mice infected at the base of the tail and footpad with high inocula of bioluminescence-expressing *L. major* parasites. In addition, we describe the first steps of a natural infection model in BALB/c mice infected with low inocula of bioluminescence-expressing *L. major* parasites in the ear pinna.

2 Materials

Carry out all procedures in the biological cabinet using aseptic technique, unless otherwise noted.

2.1 Basic Equipment and Reagents

1. Hemocytometer.
2. Tecan Infinite M200 or M1000 or similar Luminescence Plate Reader.
3. Fetal Bovine Serum (FBS): 500 mL.
4. Schneider's Drosophila Medium (Modified): 1 L (Lonza Bio Whittaker).
5. XenoLight D-Luciferin potassium salt (PerkinElmer).
6. Sodium hydroxide: 10 M Solution.
7. Distilled Water (dH_2O): 1000 mL.
8. Glycerol.

9. Phosphate buffered saline: 1× Solution.
10. Nunc MicroWell 96-well flat bottom white microplates.
11. *Leishmania* Promastigote Culture Media: Inactivate FBS in a 56 °C water bath for 40 min (*see* **Notes 1** and **2**). Combine 250 mL of heat inactivated FBS and 1 L Schneider's media. Cover and mix for 30 min on a stirrer with a stir bar, adjust pH to 7.4–7.45 using NaOH (*see* **Note 3**). Filter-sterilize (*see* **Note 4**) and store at 4 °C (*see* **Note 5**).
12. 200 × Luciferin Solution: Dissolve 1 g of XenoLight (D-Luciferin potassium salt) in 33.3 mL dH_2O (*see* **Note 6**). Store aliquots at −20 °C.
13. Promastigote Counting Solution: Mix 15 mL of glycerol with 85 mL of 1× PBS (15% glycerol solution). Vortex, filter-sterilize, and store at room temperature.

2.2 Animal In Vivo Imaging

1. Female BALB/c mice weighing 20–25 g (*see* **Notes 7–9**).
2. PerkinElmer Spectrum In Vivo Imaging System (IVIS) instrument (PerkinElmer), or similar imaging system (*see* **Note 10**).
3. V-1 Table Top Laboratory Animal Anesthesia System (VetEquip, or similar).
4. Anesthesia Induction Chamber (VetEquip, or similar).
5. Caliper instrument with 0.1 mm sensitivity for measuring lesion size.
6. Luciferin (D-luciferin potassium salt). Luciferin is dissolved in dH_2O and prepared fresh at 20 mg/mL immediately before the study starts.
7. Amphotericin B (AmBisome) is used as the positive control in all the drug discovery models of murine leishmaniasis conducted in our laboratory. AmBisome should be kept in the refrigerator at all times. Before the start of each study, the bottle contents are dissolved in 12 mL sterile dH_2O. This yields a 4 mg/mL stock solution that should be kept at 4 °C at all times and is used to freshly prepare the desired concentration of AmBisome working solution every time that the drug is administered to BALB/c mice.
8. Isoflurane, Abbot Laboratories: 250 mL (Burns Vet Supply), 2009 Caliper Life Sciences, Small Animal Anesthesia Protocol: http://www.invivoimagingsuccess.com
9. Ketamine HCl (Fort Dodge Animal Health): Dilute 5:1 ketamine in normal saline (0.9% NaCl) to working solution of 20 mg/mL. This reagent is used only during the subcutaneous (SC) infections in the footpad model (*see* **Note 11**).
10. Activated charcoal evacuation filters (Omnicon) (*see* **Note 12**).

11. 70% ethanol or other disinfectant as appropriate for each facility.
12. Nair hair removal with aloe and lanolin, purchased over-the-counter.
13. Needles (various sizes) and 0.5 mL insulin syringes for animal infections as well as luciferin injections.
14. Living Image Software (Xenogen Corporation) for assessing total photon emissions (bioluminescence signal) from the area of interest. Version 3.2 up to 4.5 were used in our laboratory (*see* **Note 13**).
15. Statistical analysis was performed using GraphPad Prism 7.04 software package (GraphPad Software, Inc.) and Microsoft Excel statistical package.
16. One-way ANOVA with Dunnett's multiple comparison test and an unpaired t-test with Welch's correction were used to compare mean lesion size and bioluminescence signal differences between group means. A p-value < 0.05 was considered statistically significant.

3 Methods

3.1 Culture Preparation for Animal Infections

Carry out all procedures in the biological cabinet using aseptic technique unless otherwise noted.

3.1.1 Reviving Frozen Leishmania Promastigote Cultures

L. major strain NIH173 (MHOM/IR/-/173) was a gift from Dr. Geneviene Milon, Department of Parasitology and Mycology, Institute Pasteur, Paris, France [17, 19, 20].

1. Transfer the entire contents of the thawed *Leishmania* cryotube to a 25 cm^2 plug-seal flask (*see* **Notes 14** and **15**).
2. Add 15 mL of *Leishmania* promastigote culture media.
3. Transfer the promastigote culture to the 27 °C incubator (*see* **Note 16**).

3.1.2 Expansion of Leishmania Promastigote Culture for Animal Infections

1. Once the *Leishmania* cultures have been fully revived from frozen, the cultures should be expanded into a larger volume (*see* **Note 17**).
2. To 150 mL of *Leishmania* promastigote culture media in a 75 cm^2 plug-seal flask, add 5 mL of revived culture (*see* **Note 18**).
3. Incubate at 27 °C and monitor parasites daily for growth.

Table 1
Dilution factors for promastigote quantification

1:2	20 μL culture + 20 μL (PBS + 15% glycerol)
1:5	20 μL culture + 80 μL (PBS + 15% glycerol)
1:10	20 μL culture + 180 μL (PBS + 15% glycerol)
1:20	20 μL culture + 380 μL (PBS + 15% glycerol)
1:30	20 μL culture + 580 μL (PBS + 15% glycerol)
1:40	20 μL culture + 780 μL (PBS + 15% glycerol)
1:50	20 μL culture + 980 μL (PBS + 15% glycerol)

3.1.3 Staging of Promastigotes Cultures (See **Note 19**)

1. For visual inspection, observe the promastigotes daily on the inverted microscope to determine when the parasites have transitioned to the metacyclic stage (*see* **Note 20**). Once the parasites have reached this point they are the most virulent and should be prepared for animal infections. For quantification methods to determine parasites stage, count the parasites using the following method (*see* **Note 21**): Mix the cultures gently and remove 0.5 mL and place in a microtube. In another microtube, dilute the parasites using promastigote counting solution (*see* **Note 22**). Use one of the dilutions listed in Table 1 (*see* **Note 23**). Vortex the diluted parasites for 15 s. Place 10 μL of the diluted parasite culture in each chamber of the hemocytometer. For each chamber, count the number of parasites in the 5 squares (labeled with an H), as shown in Fig. 1 (*see* **Note 24**). Repeat the steps daily until metacyclic stage is reached (*see* **Note 25**).
2. To determine the parasite density, use the following calculations:

$$\text{Average count} = \frac{\text{Total parasites counted in the 10 squares}}{2}$$

$$\frac{\text{Parasites}}{\text{mL}} = (\text{Average count})(5 \times 10^4) \times (\text{Dilution factor } (5,\ 10,\ \text{etc}))$$

$$\text{Total number of Parasites} = \left(\frac{\text{Parasites}}{\text{mL}}\right)(\text{Total volume})$$

3.1.4 Preparation of the Cultures for Animal Infection (See **Note 26**)

1. Transfer the cultures to 50 mL conical tubes.
2. Spin the tubes at 872 × *g* for 15 min.
3. Decant the media and resuspend in 25 mL of sterile 1× PBS (*see* **Note 27**).

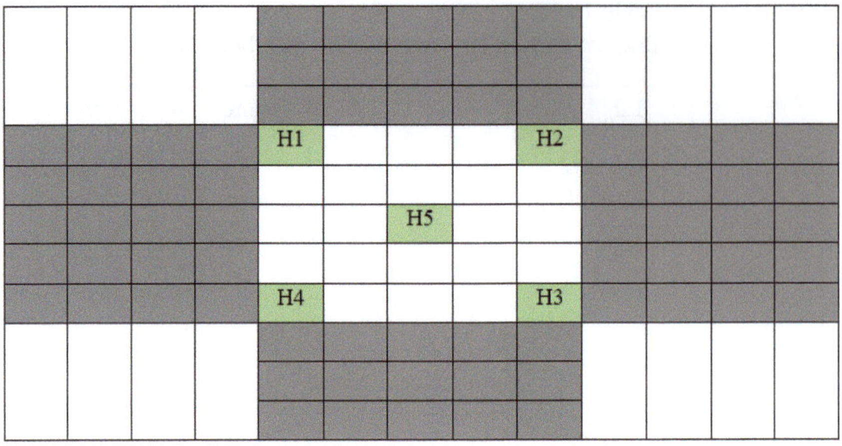

Fig. 1 Areas of the hemocytometer for counting *Leishmania* promastigotes

4. Vortex the tube(s) for 20 s to break up any pelleted promastigotes.
5. Spin again at 872 × *g* for 15 min.
6. Decant the media and resuspend each tube in 25 mL of sterile 1× PBS.
7. Spin for a third time at 872 × *g* for 15 min.
8. Decant the media and resuspend in a small volume of sterile 1× PBS (*see* **Note 28**).
9. Count the parasites as previously described on the hemocytometer.
10. Dilute the parasite in additional sterile 1× PBS to reach the desired concentration (*see* **Notes 29** and **30**).

3.1.5 Testing of Relative Luminescence of Prepared Cultures

1. Remove 100 μL from the prepared culture for animal infection and place in the well of a 96-well flat bottom white plate (*see* **Note 31**).
2. Add 10 μL of prepared 10× luciferin (*see* **Notes 32** and **33**).
3. Read the plates on a luminescence plate reader (*see* **Notes 34** and **35**).

3.2 Rodent Models of CL: Setting Up the IVIS Instrument and the Living Image Software

All *Leishmania*-infected animals should be housed in ABSL-2 conditions.

1. Look inside the imaging chamber for objects.
2. Open Living Image Software and initialize. Choose the bioluminescence option.
3. Confirm that the excitation filter is set to "Block" and the emission filter is set to "Open".

4. Set the binning and the F/Stop. Confirm the default binning (which determines the pixel size on the CCD), and F/Stop level or select a new level (*see* **Note 36**).
5. Confirm the exposure time (*see* **Note 37**).
6. Use the high throughput (5 mice) with 23 cm field of view (FOV) by using FOV D from the drop down list selection. Use FOV C, B, and A for lower throughput (less mice in the FOV) and increased resolution. The highest resolution is 20 μm and it is provided by using the FOV A.
7. Fill the vaporizer with isoflurane. Dial should be in off position for filling.
8. Turn on the oxygen tank and ensure that the pressure is approximately 55 psi. Set the oxygen flowmeter to 1 L/min. Place the mice in the induction chamber and close the lid.
9. Turn the vaporizer dial to 2.5–3%. In approximately 2–3 min when the animals are asleep, set the vaporizer dial to 1.5–2% (*see* **Note 38**).
10. Turn on the vaporizer for the IVIS machine nose cones to 2%.
11. Open the induction chamber, remove the mice, and place their noses in the nose cones inside the imaging chamber. Close the door to the imaging chamber.
12. Press "Acquire". During image acquisition, the "Acquire" button becomes a "Stop" button.
13. Fill out the label information (e.g., name of the study, group number, specific treatment information for that group, and/or is a repeat scan) (*see* **Note 39**).
14. To cancel the acquisition click "Stop".
15. When done reading all the experimental study groups, turn the vaporizer dial "Off" and then turn "Off" the oxygen tank while leaving the chamber and instrument flow switches "On" to purge the system. Once the flow drops to zero, turn "Off" the flow switches and pump.
16. Clean all surfaces and nose cones with 70% ethanol or other disinfectant as required for each facility.

3.3 Determination of Luciferin Kinetics (See Note 40)

1. Prepare 20 mg/mL luciferin (*see* **Note 41**).
2. Use five mice to conduct this study (*see* **Note 42**).
3. Manually restrain the BALB/c mice and hold them abdomen side up, and head of animal pointed downward. Use a 27 1/2 gauge needle to inject up to 200 μL D-luciferin in the lower left abdominal quadrant (intraperitoneal or IP injection) (*see* **Notes 43** and **44**).

4. Place the BALB/c mice in the induction chamber and secure the lid.
5. Turn the vaporizer dial to 2.5–3%. Animals will begin to fall asleep within 2–3 min.
6. As soon as the mice are asleep transfer them into the imagining chamber and place them on their bellies. Position their noses in the nose cones and set the vaporizer to 2% (*see* **Note 45**).
7. Set up the IVIS instrument to obtain consecutive bioluminescence readings every 2 min. Last reading should be taken 25 min post-luciferin injection (*see* **Note 46**).
8. After all images are taken, remove the animals from the imaging chamber and place them in their cages (*see* **Note 47**).
9. Select the ROI (see below), measure, and quantify the intensity of the luminescence signal for each time point.
10. Establish the optimal time post-luciferin injection for imaging *Leishmania* infection and obtaining the bioluminescence signal. The optimal time is when the bioluminescence signal is at its strongest and the wait time is the shortest (*see* **Notes 36, 48,** and **49**).

3.4 Obtaining the Bioluminescence Signal from Infected BALB/c Mice

1. Prepare 20 mg/mL luciferin.
2. Manually restrain the BALB/c mice and hold them abdomen side up, and head of animal pointed downward. Use a 27 1/2 gauge needle to inject up to 200 μL D-luciferin in the lower abdominal quadrant (intraperitoneal or IP injection). Once the luciferin is administered, place the mice in a different cage to keep track of treatments (*see* **Notes 43** and **44**).
3. Wait 5 min for the D-luciferin to be absorbed through the peritoneal tissues.
4. After the 5 min are over, place the first group of mice in the induction chamber.
5. Turn the vaporizer dial to 2.5-3 %. Animals will begin to fall asleep within 2–3 minutes.
6. Inject the second group of mice with D-luciferin (as described in Subheading 3.3) while waiting for the first group to fall asleep in the induction chamber.
7. Transfer the first group into the imaging chamber for scanning.
8. Once the 5 min are over, put the second group into the induction chamber.
9. Repeat the sequence for all groups in the study.

3.5 Data Analysis

1. Open the image of interest. An image must be active to display the area (region) of interest (ROI) tools.

2. Eliminate the saturated areas before the final measurement. Maximum counts should be set at 60000. Minimum counts should be set at 200–600 to ensure a clean image background is obtained.
3. Select "Photons/s" from the drop down menu.
4. Select measurement "ROI" from the drop down menu.
5. Specify the ROI shape: click the circle, square, or grid button (*see* **Note 50**).
6. Depending on the number of animals in each treatment group, select the number of ROI's needed.
7. Adjust the ROI size and position by dragging it over the ROI (photon intensity measurements in red, yellow, green, and/or blue). Use the mouse to resize it to cover all of the ROI.
8. Click the "Measure" button. A measurement table that contains all the needed information will be opened.
9. Click "Select All" to copy the experimental results table and paste it into a Microsoft Excel file.

3.6 Routes/Sites of Infections with Luciferase-Expressing *L. major* Promastigotes

3.6.1 Subcutaneous (SC)

This is the preferred method of infection for the footpad model.

1. Administer ketamine IP at 50 mg/kg using a 25–27 gauge needle for pain relief. Wait 5 min after injection, place a group of mice in the isoflurane induction chamber, and set isoflurane level to 2.5-3 %. Animals will fall asleep within 2–3 min. Lightly pinch a paw to test for paw withdrawal reflex. When no withdrawal is present, mice are in the surgical plane of anesthesia.
2. Use a 26–30 gauge needle to perform a subcutaneous injection into the center area of the ventral aspect of the hind foot. Maximum volume for injection at this location is 50 μL.
3. Return animals to their cages and provide external heat if necessary until they are recovered.

3.6.2 Intradermal (ID) at the Base of the Tail

This is the preferred method of infection for the base of the tail lesion suppression and lesion cure models.

1. Skin 1.5–2 cm anterior to the base of the tail of the mouse is prepared by clipping hair and then applying Nair to ensure smooth skin, at least 24 h before inoculation with parasites (*see* **Note 51**).
2. On the day of parasite injection, place a group of mice in the isoflurane induction chamber, and set isoflurane level to 2.5-3 %. Animals will fall asleep within 2–3 min. Lightly pinch a paw to test for paw withdrawal reflex. When no withdrawal is present, mice are in the surgical plane of anesthesia.

3. Use a 26–30 gauge needle to perform an intradermal injection in the prepared skin proximal to the base of the tail. Maximum volume for injection at this location is 100 μL (*see* **Note 52**). Ensure appropriate depth of intradermal injection by visualizing the needle tip between layers of skin; injected solution will form a small blister appearance.
4. Return animals to their cages and provide external heat if necessary until they are recovered.

3.6.3 Intradermal in the Ear Pinna

Intradermal infection (ID) in the ear pinna is slightly different compared to the base of the tail.

1. Place a group of mice in the isoflurane induction chamber, and set isoflurane level to 2.5–3%. Animals will fall asleep within 2–3 min. Lightly pinch a paw to test for paw withdrawal reflex. When no withdrawal is present, mice are in the surgical plane of anesthesia.
2. Use a 27 1/2–30 gauge needle to perform an intradermal injection into the ear pinna (*see* **Note 53**). Ensure appropriate depth of the intradermal injection by visualizing the needle tip between the skin and the cartilage of the ear pinna (*see* **Note 54**). Maximum volume for injection in this location is 10 μL. The mouse ear pinna can be supported for injection by taping against a plastic tube or held gently by non-crushing tissue forceps [23].
3. Return animals to their cages and provide external heat if necessary until they are recovered.

3.7 Prescreen: Lesion Suppression Model in BALB/c Mice

This model is conducted as described in Caridha et al., 2017 [17]. Percent (%) vehicle control (VC) bioluminescence signal reduction is the main experimental endpoint in this model.

1. One to 2 days prior to the infections, shave the BALB/c mice and remove the hair by applying Nair to the area around the base of the tail. Thoroughly rinse with a wet gauze to prevent skin burning and then dry the animal with dry gauze.
2. Day 0: Infect mice ID at the base of the tail with 1×10^7 stationary (or 1×10^6 metacyclic) phase luciferase-expressing *L. major* promastigotes (*see* **Note 55**).
3. Day 3 Postinfection: Determine baseline bioluminescence signal by randomly choosing 10-15 infected BALB/c mice and assessing the intensity of bioluminescence signal emitted from the infected site. Administer the first dose of the drug only if/when bioluminescence signal is present in all BALB/c mice.
4. Starting on day 3 postinfection administer drugs including the negative and positive controls, according to the study design (*see* **Notes 56–58**).

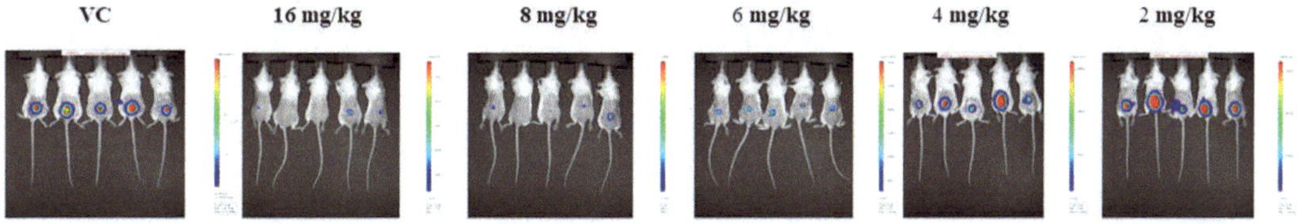

Fig. 2 Amphotericin B deoxycholate reduces the bioluminescence signal (which represents the parasite load) in a dose-response manner in the lesion suppression model conducted in BALB/c mice infected ID with 1×10^7 stationary phase *L. major* promastigotes. Percent (%) VC reduction of in vivo bioluminescence signal in BALB/c mice treated with various doses of amphotericin B deoxycholate was used to assess drug efficacy in the lesion suppression model. Mice were infected ID with 1×10^7 stationary-phase luciferase-expressing *L. major* parasites at the base of the tail on day 0. Drug treatments (2, 4, 6, 8, and 16 mg/kg amphotericin B deoxycholate, 25 mg/kg AmBisome, as well as the VC (double distilled sterile water)) were initiated on day 3 post-animal infection. All drugs and the VC were administered intraperitoneally (IP). The bioluminescence signal (which represents the light intensity over the body surface area) was measured on day 13 postinfection with the purpose of assessing the % VC parasite load suppression of different doses of amphotericin B deoxycholate at the infection site. Red and bright yellow represent the most intense signal, followed by green and then blue, which represents the weakest signal. The bioluminescence signal was detected at the infection site in all groups starting from day 0 postinfection and continued to be present in the VC group until the end of the study. The lack of a bioluminescence signal at the infection site suggested that the parasite load was nonexistent or below the limit of detection. An exposure time of 1 min and the "auto" option was used to obtain the in vivo bioluminescence images from all treatment groups but only the images taken after the 1 min exposure time are shown in the figure. The "auto" reading option was used to assess the amphotericin B deoxycholate parasite load suppression at the infection site

5. In order to accommodate treatments that might be longer or shorter (7–14 days, or even longer), the bioluminescence signal is obtained in all BALB/c mice belonging to all treatment and control groups on the day post end of treatment (Fig. 2).
6. Assess % VC mean bioluminescence signal suppression in all drug-treated groups. (*see* **Notes 59** and **60**).
7. Euthanize all BALB/c mice immediately after the % VC mean bioluminescence signal suppression for the drug - and control-treated groups are measured. Alternatively, all groups can be observed until the bioluminescence signal increases above the limit of detection and is visible and measurable in IVIS for the AmBisome-treated group. Compare the time of signal relapse in the drug treated with the timing of the relapse in the AmBisome-treated group (*see* **Notes 61** and **62**).
8. Pinpoint the most efficacious antileishmanial drugs (e.g., the ones that suppress the mean bioluminescence signal ≥50% compared to that of the VC group). Further test these compounds for efficacy in the base of the tail lesion cure model in BALB/c mice.

3.8 Base of the Tail Lesion Cure Model (See Notes 63 and 64)

This model is conducted as described in Caridha et al., 2017 [17]. Percent (%) VC lesion size reduction and/or lesion cure is the main experimental endpoint in this model. Still,

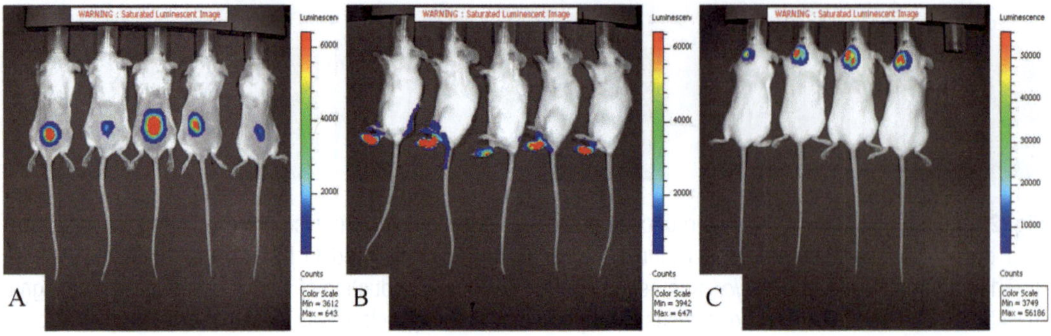

Fig. 3 Bioluminescent images of BALB/c mice infected respectively at the base of the tail, footpad, and ear pinna with (**a**) 1×10^7, (**b**) 1×10^7, and (**c**) 2×10^3 luciferase-expressing stationary phase *L. major* promastigotes. Images were taken using the IVIS Spectrum instrument and analyzed using the Living Image Software 3.2 to 4.5 respectively 20, 20, and 40 days post-animal infection

bioluminescence signal at the infection site can be used as a strong indicator for the absence or presence of parasites, as well as the tendency of the parasite load to increase, decrease, or stay unchanged (Fig. 3a).

1. Prepare the mice for infections as described in Subheading 3.6.2.
2. Infect BALB/c mice ID with 1×10^7 stationary phase (or 1×10^6 metacyclic phase) luciferase-expressing promastigotes at the base of the tail, as described in Subheading 3.6.2.
3. Shave the BALB/c mice twice a week and check for papule and/or small lesion formation.
4. Starting from the second or third week postinfection (depending on the virulence of the *Leishmania* spp. strain and the timing of lesion formation), use a caliper instrument with 0.1 mm sensitivity to measure the lesion induration diameters (length = D_1 and width = D_2). Calculate the lesion size area using the $\pi R_1 \times R_2$ formula (where $R_1 = D_{1/2}$ and $R_2 = D_{2/2}$) (*see* **Note 65**).
5. When average lesion size area reaches approximately 15–20 mm^2, assign BALB/c mice in groups of five animals such as the average lesion size of each group, including the treatment, positive, and negative control groups, are not statistically different from each other (ANOVA).
6. Before drug treatments start, determine the baseline bioluminescence signal by randomly choosing 10-15 BALB/c mice and assessing the intensity of bioluminescence signal emitted from the infected site.
7. Give drug treatments according to the study design (*see* **Note 66**).

8. On the day after the end of treatment, measure bioluminescence signal in all treatment groups including the positive and negative control groups. Calculate % VC bioluminescence signal suppression.
9. Continue to measure the lesion size and follow the intensity of the bioluminescence signal emitted by the lesion and the area surrounding it at least once a week for all treatment groups until the end of the study. In this particular model, the intensity of the bioluminescence does not correlate well with the parasite load at the base of the tail, but it can be a useful indicator of the tendency of the parasite load (increase, decrease, or stay unchanged) [17]. Additionally, this can provide useful information about the potential mode of action of the antileishmanial drug.
10. When all AmBisome treated BALB/c mice are cured (this usually happens 10–20 days after the end of the 10 day AmBisome treatments), assess % VC lesion size reduction (which is the main experimental endpoint for the base of the tail lesion cure model). Calculate % VC lesion size reduction for all study groups and determine drug efficacy (*see* **Note 67**).
11. Watch for relapse of the presence of the bioluminescence signal at the infection site in the previously healed BALB/c mice as it is a clear sign of the relapse of the disease (lesions). *Important*: In BALB/c mice, the size of the lesions at the base of the tail will increase until animals ultimately die from metastasized leishmaniasis disease. In order to prevent that, animals should be euthanized when the lesions become ≥150–200 mm^2 or if other signs of the disease, such as weight loss, are present (*see* **Note 68**).

3.9 Footpad Thickness Suppression Model

This model is conducted as described in Sacks and Melby, 2001 and 2015 [21, 22] with the one difference being that the footpad thickness was the only parameter followed throughout the study to determine drug efficacy. Reduction of the footpad thickness is the main experimental endpoint in this particular model. The parasite load in the infected footpads is followed through the measurement of in vivo bioluminescence signal during the entire disease progression in all treatment groups. It is used as a strong indicator for the presence or absence of parasites, as well as the tendency of the parasite load to increase, decrease, or stay unchanged (Fig. 3b).

1. BALB/c mice are infected SC in the hind footpad with 50 μL culture containing 1×10^7 luciferase-expressing stationary phase (or 1×10^6 metacyclic phase) promastigotes. A second VC group (VC 2) contains five mice in which footpads are injected with 50 μL parasite media. This injection is described in Subheading 3.6.1.

2. After infections, measure the footpad thickness of the infected and control uninfected footpads at least twice a week (*see* **Note 69**). Drug treatment should start as soon as the mean footpad thickness of infected footpads becomes statistically different from the mean footpad thickness of the VC 2 group (*see* **Note 70**).
3. Before the drug treatment starts, randomly choose 10-15 BALB/c mice and assess the baseline intensity of bioluminescence signal emitted from the infected footpads.
4. Administer drug treatments according to the study design (*see* **Note 66**).
5. On the day after the end of treatment, measure the bioluminescence signal in all treatment groups including the positive and negative control groups. Calculate % VC1 bioluminescence signal suppression (*see* **Note 71**).
6. Continue to measure footpad thickness in all study groups twice a week. When the mean footpad thickness of the AmBisome-treated group is not statistically different from the VC2 group (this will happen approximately 10–12 days after the end of the 10 day AmBisome treatments), assess % VC1 footpad thickness reduction (which is the main experimental endpoint for the footpad model) and determine drug efficacy (*see* **Note 72**).
7. Continue to follow the intensity of the bioluminescence signal emitted by the lesion and the area surrounding it at least once a week for all treatment groups. As stated in the base of the tail lesion cure model mentioned above, in this particular model there is not a strong correlation between the intensity of the bioluminescence signal and the parasite load at the infected footpads [17] (*see* **Note 73**).

3.10 Ear Infection Model [12, 23] (Fig. 3c)

3.10.1 Sand Fly Feeding Assay/Ear Infection Model

Phlebotomine sand fly rearing is described extensively in literature [24], so will not be discussed here. The use of phlebotomine sand fly salivary components have been useful to the study of CL. This particular model utilizes sand fly feeding behavior and saliva combined with known numbers of *Leishmania* spp. parasites.

1. Twelve hours prior to the start of each sand fly feeding assay, gently aspirate 10 sand flies into a dram vial with a mesh screen over the opening, through which the flies can feed. One vial of 10 flies will be used to expose a single mouse.
2. The sand flies should be sugar-starved for 12 h prior to exposure of the mouse to ensure hunger.
3. Anesthetize BALB/c mice with injected ketamine–xylazine and expose the mouse ear to sand flies as described in literature [12], *see* Fig. 4 (*see* **Note 74**).

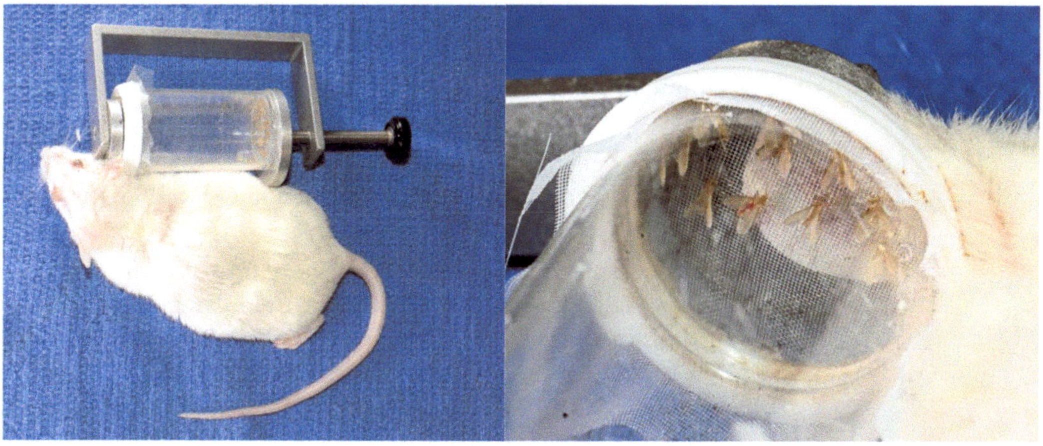

Fig. 4 Example of setup for sand fly feeding experiment on mouse ear with vial of 10 flies clamped to anesthetized mouse's ear (left), and flies probing and feeding on mouse ear (right); one can see the fed, blood-filled fly (center of image)

4. Immediately after the flies have fed on the mouse ear, infect mice ID with 10 μL culture containing 2×10^4 stationary phase (or approximately 2×10^3 metacyclic phase) luciferase-expressing *L. major* promastigotes into the ear pinna that was exposed to the fly bites, as described in Subheading 3.6.3. The negative control group for this assay consists of mice exposed to sand flies feeding on their ear and then injected ID with culture media instead of parasites.
5. Provide external heat to the mice as they recover from anesthesia (*see* **Note 75**).
6. Starting from the second week postinfection (depending on the virulence of the *Leishmania* spp. strain and the timing of lesion formation), check for papules/lesions twice each week, and use a caliper instrument with 0.1 mm sensitivity to measure the lesion induration diameters (length = D_1 and width = D_2). Calculate the lesion size area using the $\pi R_1 \times R_2$ formula (where $R_1 = D_{1/2}$ and $R_2 = D_{2/2}$) (*see* **Note 61**).
7. Bioluminescence readings on all mice can be taken on the same days as caliper measurements (*see* **Note 76**).

3.10.2 Ear Infection Model With or Without Salivary Gland Sonicate (SGS)

This model is used extensively in literature as a site that requires lower doses of *Leishmania* spp. parasites to produce lesions. The mouse pinna can be infected with parasites alone or a combination of parasites and phlebotomine sand fly salivary gland sonicate (SGS). This model is conducted as described in Cawlfield et al., 2018 [12], allowing the option of infecting mice with *Leishmania* culture with or without phlebotomine SGS.

1. From a sand fly colony, aspirate newly emerged adults into smaller cups and hold for 7 days.
2. Immobilize the flies by aspirating them into soapy water, and run soapy water over fine mesh screen to isolate the flies. Rinse the flies with deionized water.
3. Place flies on microscope slide containing 1× PBS and place under dissecting microscope.
4. Using dissecting pins, separate the head from each fly, and tease salivary glands out of the back of the head. Each pair of salivary glands can be used to infect 5 mice (0.2 pair/mouse) (*see* **Note 77**).
5. Glands are disrupted by ultrasonication for 40 cycles, each lasting approximately 2 min, creating SGS.
6. Immediately place tube into storage at −80 °C until use.
7. Prior to the start of inoculations, remove the tube from freezer and place on ice.
8. Combine 4 μL of SGS (0.2 pairs of glands) with 6 μL *Leishmania* spp. culture containing 2 × 10^4 stationary phase (or 2 × 10^3 metacyclic phase) luciferase-expressing promastigotes and inject into the ear pinna of an anesthetized mouse as described in Subheading 3.6.3. The control group of animals will be injected with 4 μL of SGS combined with 6 μL of 1× PBS.

4 Notes

1. Thaw FBS overnight in a 4 °C refrigerator.
2. After completion of heat inactivation, the FBS should be allowed to cool back down to room temperature and then can be aliquoted into 50 mL conical tubes and stored at −20 °C freezer, or used immediately for preparation of media.
3. Calibrate the pH meter while the media is mixing.
4. Due to the amount of FBS, connecting the filter unit to a vacuum will not work to filter-sterilize the media. Instead it is better to allow gravity to filter the media overnight at 4 °C. To do this, open only the top portion of the bag that the filter unit came in and add the media to the filter unit. Secure the lid and tape the bag close. The media should filter through completely within 24 h.
5. Media is good for up to 1 month after preparation.
6. Due to the light sensitivity of luciferin, preparation of luciferin should be conducted without the lights on in the lab or the biological cabinet. Since the vial will not hold all 33.3 mL of

dH_2O, add 10 mL of dH_2O, seal, and vortex for 20 s. Transfer the contents to a 50 mL conical tube. Repeat this two more times and then add in the remaining 3.3 mL to bring the final volume up to 33.3 mL.

7. All animals were purchased from the same vendor to prevent possible genetic variations by using animals from different colonies that may interfere with the results. Facilities were kept under 12 h light–dark period cycles, at a room temperature of 22 ± 2 °C, 40–70% humidity, and air-conditioned with 20 air changes per hour. Filtered water and pelleted diet were provided ad libitum. Animal husbandry and room/cage cleaning and sanitation was provided in accordance with WRAIR/NMRC SOP No. 555 entitled "Husbandry of Rodents."
8. Ethics statement: The animal protocol for this study was approved by the Walter Reed Army Institute of Research, Institutional Animal Care and Use Committee (Protocol Number 16-ET-33) in accordance with national and Department of Defense guidelines. Research was conducted in an AAALACi accredited facility in compliance with the Animal Welfare Act and other federal statutes and regulations relating to animals and experiments involving animals and adhere to principles stated in the Guide for Care and Use of Laboratory Animals, NRC Publication, 2011 edition.
9. Other mouse strains, such as C57BL/6, CBA, CsS-16, DBA/2, CBA/H, and C3H/HeN, as well as other virulent *Leishmania* spp. parasites can be used in the mouse models of CL [4, 25].
10. While all the methodologies and methods described here were conducted using the IVIS Spectrum (PerkinElmer), other instruments, such as IVIS Lumina, and PET/CT imaging instruments, can be successfully used to monitor bioluminescence emission and disease progression in experimental models of murine *Leishmania*.
11. Ketamine–HCl is administered to alleviate pain while performing SC infections in the footpads, but not during the ID infections at the base of the tail and ear pinna.
12. *Caution*: Activated charcoal evacuation filters (canisters) are used to absorb isoflurane vapors that would otherwise vent out. Canisters should be weighed before use 1–2 times a week (this depends on the frequency and length of using the IVIS instrument). Canisters should be replaced before the weight increase reaches 50 g.
13. The most recent version of the Living Image Software is 4.5.5. Using the most up to date version of the Living Image Software is recommended.

14. Parasites should be revived from frozen approximately 10 days in advance of the animal infection. This time frame will vary depending upon the species and growth rate for that particular parasite strain.
15. Frozen promastigote cultures should be stored in the liquid nitrogen tank and can be thawed in room temperature water.
16. Monitor the parasites daily on an inverted scope to check for growth/revival of the culture.
17. Cultures being used for animal infections should be used as soon as possible after reviving to limit the chance of reducing virulence.
18. Multiple flasks may need to be prepared depending upon the number of animals to infect, and parasites/mL need for each infection.
19. Parasite stage can be determined by either visual inspection with an inverted scope or by daily quantification of the parasites.
20. At the start of the growth cycle, the promastigotes will appear more round in shape with a less-visible tail. As the promastigotes progress in the growth cycle and enter late log stage the parasites will start to elongate. Upon reaching metacyclic stage, the parasites will be extremely elongated and thinner than previously observed.
21. Parasites should still be monitored on the inverted scope to check for overall appearance of the culture. Cultures should be checked to ensure there is no contamination and that there are few rosettes present.
22. The promastigote counting solution will slow down the mobility of the parasites, making it easier to count them.
23. Dilution factor used is determined by the density of the parasites. Ideally, 50–60 parasites should be counted in total for both sides of the hemocytometer. Adjust the dilution as needed to reach this goal.
24. Count the parasites that are inside of each square and the ones that are touching the lines only on the top and left side of the square. Parasites touching the right and bottom lines should not be counted. If rosettes are present, count them as one parasite.
25. For the first few days the parasites should be growing exponentially (log phase). The parasites will then slow down and the count should remain stable for 1–2 days. This is the point when the promastigotes have reached metacyclic stage and should be used for animal infections. Cultures that are left beyond this stage will begin to drop in density which is an indicator that the culture has entered stationary phase.

26. Although we prepare the cultures when the majority of the promastigotes are in the metacyclic stage of the life cycle, this is considered a stationary phase culture preparation since we do not select for only metacyclic stage promastigotes to be used for the animal infections through methods such as a Percoll gradient. As a result, the final preparation of culture will be comprised of both metacyclic and stationary stage promastigotes.
27. The number of 50 mL conical tubes can be reduced at this point by resuspending multiple tubes in the same 25 mL of 1× PBS.
28. For the final suspension, use approximately 1 mL of 1× PBS per every 30 mL of culture that was originally spun down. This will vary depending upon the strain and final desired concentration of the parasites.
29. Typically parasites are diluted to 1×10^8 parasites/mL. Depending upon the rodent model being used, volume for infection, and *Leishmania* spp., the concentration may increase or decrease.
30. If the parasites are too dilute, concentration the parasites by spinning at 872 × *g* for 15 min and resuspending in a smaller volume. Count the parasites again and dilute to the appropriate concentration.
31. Preferably run at least two wells for testing, placing 100 μL in each well.
32. To test for luminescence of a prepared culture, a 10× luciferin solution is needed. This solution should be prepared fresh right before using. No more than 2 mL total will be needed to test the parasites. Using the previously prepared 200× luciferin solution, prepare a 10× solution by mixing 1.9 mL of 1× PBS with 0.1 mL of 200× luciferin. Solution should be kept in the dark until added to the plate.
33. Final luciferin concentration in the well is 150 μg/mL.
34. Plates can be read immediately on the reader or incubated for 15 min in the dark at room temperature. No difference in RLUs was observed between these two time points. If the plate is incubated longer than 30 min, RLU activity will start to decrease.
35. If the RLU reads pass the cut off value for infection, which is dependent on the species, the parasites should be used immediately for animal infections. If the RLU reads do not pass the cutoff value, cultures should be discarded and new parasites brought up from frozen for use.
36. Once selected, the binning and F/Stop should be the same throughout the study.

37. Use of auto exposure (which automatically sets the exposure time, F/Stop, and binning to keep the signal within an optimal target range) is highly recommended. If chosen, auto exposure should be used to obtain the bioluminescence signal throughout the study.

38. Keep the isoflurane flow as low as possible, to ensure that the mice remain asleep until they are transferred to the imagining chamber and that the leak of isoflurane vapors from the induction chamber is minimal. Apply eye lubrication as soon as mice are transferred to the nose cones.

39. If the bioluminescence signal at the infection site is strong in some, but below the limit of detection for one or several other mice belonging to the same treatment group, there is a possibility that an error in luciferin injection might have occurred (e.g., injections may have hit abdominal organs, which limits the luciferin absorption through the peritoneal tissues). In these cases, another dose of luciferin should be administered and bioluminescence signal should be measured again.

40. Assessing the luciferin kinetics (optimal settings on the IVIS instrument and the right time after luciferin injections when the bioluminescence signal is strong but the exposure time is short) is a very important first step that should be performed while validating the in vivo models of murine CL.

41. The dose of D-luciferin should be such that it minimizes the amount of this expensive reagent that is used in these studies and maximizes the bioluminescence signal emitted by the infection site. 15–20 mg/mL is a reasonable concentration to use in the in vivo models of CL as working solution of D-luciferin [16, 17].

42. In order to offset variability between bioluminescence signals emitted from different mice, at least 4–5 mice should be used for the kinetics study.

43. The needle should penetrate just through the abdominal wall in order to avoid D-luciferin being injected in different organs.

44. Final dose of luciferin should be 150–200 mg/kg [16, 17].

45. BALB/c mice should rest on their bellies when they are infected at the base of the tail. When infected in the footpad or ear pinna, they should be positioned slightly tilted on their sides, such as an unobstructed view of the infected site is clearly visible.

46. Do not keep animals in the IVIS chamber for more than 25–30 min.

47. Place animals on their backs to facilitate breathing. Animals should be awake and start moving around the cage within several minutes. Do not leave the animal room before all animals are awake and active in their cages.

48. In murine models of experimental *Leishmania*, bioluminescence signal usually reaches a plateau 15–20 min post-D-luciferin injection [16, 17].
49. While obtaining the bioluminescence signal, increase in exposure time will not necessarily yield higher light intensity. Pay attention to the warning windows about presence of saturated images. Choose Display and then Saturation Map to visualize the saturated pixels within the image. Readjust sensitivity settings to eliminate areas of saturation before the final measurement.
50. In the murine models of *Leishmania* the round shape is used to determine the ROI. If needed, the round shape can be elongated to form an oval shape as well.
51. Nair can irritate skin. Thoroughly clean mouse skin after application of Nair and allow at least 24 h to ensure no confounding irritation has occurred.
52. Up to 200 μL have been injected subcutaneously at this location.
53. Either surface of ear pinna can be inoculated, but the concave surface is most consistent for ID depth; mice smaller than 18 g require closer to 30 gauge needles.
54. If needle goes through cartilage, parasite ulcer may form on both sides of ear.
55. In all models described in this chapter, BALB/c mice can be infected with stationary or metacyclic promastigotes. The WRAIR team did not find a significant difference in disease progression and/or timing of lesion formation in BALB/c mice infected with stationary vs. metacyclic luminescence-expressing *L. major* promastigotes. In several other sources of literature, use of metacyclic promastigotes is strongly suggested. Methodology for the preparation of metacyclic promastigotes can be found here [23, 26].
56. Drugs are typically administered once a day IP in order to increase drug bioavailability. Based on their properties, drugs can be given orally (PO), intravenously (IV), subcutaneously (SC), or intramuscular (IM). Drugs with a very short half-life can be given twice (BID), or even three times a day (TID).
57. The negative control group is dosed with the solvent used for drug formulations. WRAIR team mainly uses HECT (hydroxyethyl cellulose + 0.2% Tween 80) as a drug solvent, but 1× PBS, dH_2O, and other solvents can be used as well. More information about vehicle selection and solute preparation can be found in literature [27].
58. AmBisome is used as the positive control in the mouse lesion suppression screen and all the other drug discovery models of murine *Leishmania*. AmBisome is administered once a day IP at 25 mg/kg.

59. During the first 18–20 days after infection with the luciferase-expressing *L. major*, the intensity of the bioluminescence signal in the VC group increases exponentially and there is a very strong correlation between the bioluminescence signal and the parasite load at the infection site. As a result, a significant decrease in the bioluminescence signal can be used for assessing the antileishmanial drug efficacy in the BALB/c mouse/ *L. major* lesion suppression model. In addition, WRAIR team has shown that there is a very good correlation ($R^2 = 0.80$) between % VC suppression of the mean bioluminescence signal on the day after the end of the treatment with the % reduction of the lesion sizes on day 60 postinfection (assuming the mice are kept that long) [17].
60. It might be useful in drug efficacy studies, in which drug treatment lasts for $\geq$10 days, to assess the bioluminescence signal in all treatment groups after half of the drug doses are given (e.g., after the fifth dose is given in a once a day, 10 consecutive days treatment regimen). It is also useful to assess the bioluminescence signal at any time during the treatment that makes sense for the particular drug in use and its known/ supposed mechanism of action. This gives an additional opportunity to shed some light on the timing in which parasite load starts to diminish compared to the AmBisome-treated group.
61. AmBisome, when given for 10 consecutive days (QD, IP) at 25 mg/kg, reduces the bioluminescence signal below the limit of detection in BALB/c mice. *Caution*: Increasing the dose of AmBisome to 37.5 and 50 mg/kg is not recommended since significant weight loss, other signs of toxicity, and/or animal death can occur [17].
62. Not all BALB/c mice belonging to either AmBisome- or drug-treated groups will relapse at the same time.
63. In the base of the tail lesion cure model, the intensity of the bioluminescence signal reduction emitted from the lesion and the surrounding area is not an accurate endpoint for determining parasite load reduction and drug efficacy. The presence of dark scabs and necrotic tissue quenches the amount of bioluminescence signal emitted by the affected area. Despite this, measurement of the bioluminescence signal around the lesion area can be used to assess the tendency of the parasite load to increase or decrease (such as the case of AmBisome treatment in which case the bioluminescence signal is reduced below the limit of detection by the end of the 10 day, 25 mg/kg IP treatment). In addition, in the case in which lesions are cured (such as in the AmBisome and some very efficacious antileishmanial compounds) relapse can be measured by the timing after which the *Leishmania* parasites become visible again (bioluminescence signal increase above the limit of detection).

64. One hundred percent (100%) suppression (lesion cure) was considered 100% re-epithelialization of the lesion (lesion size, 0 by 0 mm^2).
65. Based on the fact that *Leishmania* spp. caused lesions that are not round and symmetrical, we calculate the lesion surface area by using the $\pi R_1 \times R_2$ formula (as described above). While in our opinion this is the most accurate way of measuring the surface area, many laboratories follow the lesion diameter as the indicator of the disease progression and/or drug efficacy.
66. As in the other screens, drugs can be given IP, IV, PO, IM, or SC. Length (days) and frequency (QD, BID, or TID) of treatment can vary as well.
67. Usually the bioluminescence signal at the infection site falls below the limit of detection in all BALB/c mice belonging to the AmBisome positive control group toward the end of the 10 day treatment and stays below that limit while the lesions heal. In most cases, the parasite load in BALB/c mice belonging to the AmBisome group grows above the limit of detection (relapses) 1.5–3 weeks after lesions completely heal. The reappearance of bioluminescence signal (parasites) at the lesion site means that the healed lesion will reopen soon.
68. In most studies, expect to euthanize BALB/c mice belonging to the VC and other groups treated with ineffective drugs well before the mice that belong to the AmBisome and other groups treated with efficacious antileishmanial drugs.
69. When infected with a large parasite inoculum, footpad thicknesses will grow fast. It is of great importance that the drug treatments start as soon as the footpad thickness in the infected animals is significantly different from the uninfected ones.
70. In all BALB/c mouse models of CL, parasites metastasize from the infection site to the lymph nodes, bone marrow, spleen, and liver which makes these models very stringent [4, 28]. For this reason, the sooner the drug treatments start, the better it is for assessing drug efficacy of potential antileishmanial compounds. Use of 3D reconstruction of bioluminescence using DLIT can provide useful information about the visceralization of luciferase-expressing *L. major* parasites in draining lymph nodes and possibly in other internal organs as well (Fig. 5). Images in Fig. 5 were taken using Living Image Software 3.2. For a step to step guide to perform 3D reconstruction of bioluminescence using DLIT go to: http://www.perkinelmer.com/Content/LST_Software_Downloads/Living%20Image%2032%20User%20Manual_PN125112-8234.pdf
71. Assessment of bioluminescence signal at the infection site can help adjust/determine the length of drug treatment. If bioluminescence signal at the infection site (base of the tail, footpad, and/or ear pinna) continues to decrease and if toxicity is not an

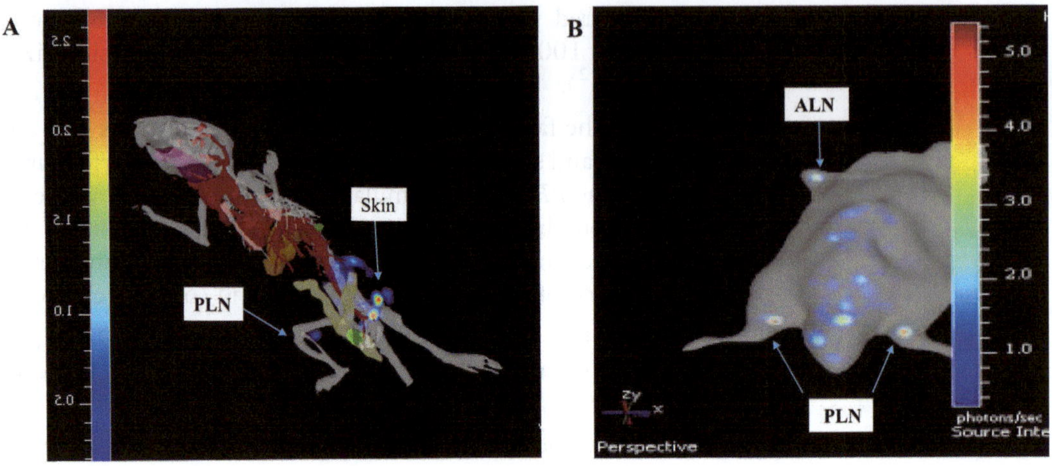

Fig. 5 Three dimensional reconstruction of bioluminescence using DLIT in a BALB/c mouse on (**a**) 30 and (**b**) 50 days postinfection at the base of the tail with 1×10^7 stationary phase luciferase-expressing *L. major* parasites. Photon intensity measurements in the brighter red areas of the image are larger than those in the dimmer green-blue areas. Thirty days postinfection (**a**) the *L. major* parasites are mostly located in the infected skin at the site of infection. In addition, the parasites have migrated to the draining popliteal lymph node (PLN). At 50 days postinfection (**b**) there is an increased parasites load of the *L. major* parasites in the skin around the infection site. Moreover, the parasites are present in large quantities in both the popliteal and axillary lymph nodes (ALN) and possibly in the internal organs

issue, the drug can be given longer than previously planned. *Important*: Even if experimental drugs are given for longer periods of time, only 10 consecutive doses of the positive control AmBisome should be administered. Most of the known antileishmanial drugs in humans are given for more than 10 days.

72. The study should be wrapped up before ulcerated sores, extensive necrosis, and dermal erosion are present in the VC or other groups treated with ineffective antileishmanial drugs.

73. As in the lesion cure model, assessing the tendency of the bioluminescence signal (parasite load) to increase, decrease, or stay unchanged, can be a useful tool to assess drug efficacy and a possible mode of action of potential antileishmanial drugs. In addition, as described above, following all "cured" groups for a possible relapse (reappearance of bioluminescent signal in the infected footpad) and comparing it with the timing of the relapse in the AmBisome-treated group can be a useful tool to assess drug efficacy.

74. When exposing to sand flies, inhaled anesthetics cannot be used for the mice, as these agents will also anesthetize the flies. Flies frequently probe without taking a meal, but they will still expose the area of skin to their saliva. Allow the flies to feed undisturbed for either 30 minutes, or until noted that all flies have fed.

75. Injectable anesthesia at doses required to keep mice still for the 30 min of fly feeding can last 45 min or longer; for this reason mice will need a heating pad or lamp to keep warm.
76. Bioluminescence signal was measurable 1–2 weeks before external lesion development.
77. Twenty (20) pairs of salivary glands in 400 μL of 1× PBS were used.

Acknowledgments

Disclaimer: Material has been reviewed by the Walter Reed Army Institute of Research. There is no objection to its presentation and/or publication. The opinions or assertions contained herein are the private views of the author, and are not to be construed as official, or as reflecting true views of the Department of the Army or the Department of Defense.

Research was conducted under an approved animal use protocol in an AAALACi accredited facility in compliance with the Animal Welfare Act and other federal statutes and regulations relating to animals and experiments involving animals and adheres to principles stated in the *Guide for the Care and Use of Laboratory Animals*, NRC Publication, 2011 edition.

References

1. World Health Organization (2010) Control of the leishmaniases. World Health Organ Tech Rep Ser 949:xii–xiii 1-186
2. Croft SL, Seifert K, Yardley V (2006) Current scenario of drug development for leishmaniasis. Indian J Med Res 123(3):399–410
3. Pigott DM, Bhatt S, Golding N et al (2014) Global distribution maps of the leishmaniases. Elife 3. https://doi.org/10.7554/eLife.02851
4. Mears ER, Modabber F, Don R, Johnson GE (2015) A review: the current in vivo models for the discovery and utility of new anti-leishmanial drugs targeting cutaneous leishmaniasis. PLoS Negl Trop Dis 9(9): e0003889. https://doi.org/10.1371/journal.pntd.0003889
5. Robledo SM, Carrillo LM, Daza A et al (2012) Cutaneous leishmaniasis in the dorsal skin of hamsters: a useful model for the screening of antileishmanial drugs. J Vis Exp 62:3533. https://doi.org/10.3791/3533
6. Milon G, Del Giudice G, Louis JA (1995) Immunobiology of experimental cutaneous leishmaniasis. Parasitol Today 11(7):244–247
7. Gomes-Silva A, Valverde JG, Ribeiro-Romao RP et al (2013) Golden hamster (*Mesocricetus auratus*) as an experimental model for *Leishmania* (*Viannia*) *braziliensis* infection. Parasitology 140(6):771–779. https://doi.org/10.1017/S0031182012002156
8. Ribeiro-Romao RP, Moreira OC, Osorio EY et al (2014) Comparative evaluation of lesion development, tissue damage, and cytokine expression in golden hamsters (*Mesocricetus auratus*) infected by inocula with different *Leishmania* (*Viannia*) *braziliensis* concentrations. Infect Immun 82 (12):5203–5213. https://doi.org/10.1128/IAI.02083-14
9. Kimblin N, Peters N, Debrabant A et al (2008) Quantification of the infectious dose of *Leishmania major* transmitted to the skin by single sand flies. Proc Natl Acad Sci U S A 105 (29):10125–10130. https://doi.org/10.1073/pnas.0802331105
10. Abdeladhim M, Kamhawi S, Valenzuela JG (2014) What's behind a sand fly bite? The profound effect of sand fly saliva on host hemostasis, inflammation and immunity. Infect Genet

Evol 28:691–703. https://doi.org/10.1016/j.meegid.2014.07.028

11. Belkaid Y, Kamhawi S, Modi G et al (1998) Development of a natural model of cutaneous leishmaniasis: powerful effects of vector saliva and saliva preexposure on the long-term outcome of *Leishmania major* infection in the mouse ear dermis. J Exp Med 188 (10):1941–1953
12. Cawlfield A, Vesely B, Ngundam F et al (2018) Use of in vivo imaging system technology in *Leishmania major* BALB/c mouse ear infection studies. J Med Entomol 55(2):429–435. https://doi.org/10.1093/jme/tjx219
13. Greer LF 3rd, Szalay AA (2002) Imaging of light emission from the expression of luciferases in living cells and organisms: a review. Luminescence 17(1):43–74. https://doi.org/10.1002/bio.676
14. Lang T, Goyard S, Lebastard M, Milon G (2005) Bioluminescent *Leishmania* expressing luciferase for rapid and high throughput screening of drugs acting on amastigote-harbouring macrophages and for quantitative real-time monitoring of parasitism features in living mice. Cell Microbiol 7(3):383–392. https://doi.org/10.1111/j.1462-5822.2004.00468
15. Roy G, Dumas C, Sereno D et al (2000) Episomal and stable expression of the luciferase reporter gene for quantifying *Leishmania spp.* infections in macrophages and in animal models. Mol Biochem Parasitol 110(2):195–206
16. Thalhofer CJ, Graff JW, Love-Homan L et al (2010) In vivo imaging of transgenic *Leishmania* parasites in a live host. J Vis Exp 41:1980. https://doi.org/10.3791/1980
17. Caridha D, Parriot S, Hudson TH et al (2017) Use of optical imaging technology in the validation of a new, rapid, cost-effective drug screen as part of a tiered in vivo screening paradigm for development of drugs to treat cutaneous leishmaniasis. Antimicrob Agents Chemother 61(4):e02048-16. https://doi.org/10.1128/AAC.02048-16
18. Schuster S, Hartley MA, Tacchini-Cottier F, Ronet C (2014) A scoring method to standardize lesion monitoring following intra-dermal infection of *Leishmania* parasites in the murine ear. Front Cell Infect Microbiol 4:67. https://doi.org/10.3389/fcimb.2014.00067
19. Lecoeur H, Buffet PA, Milon G, Lang T (2010) Early curative applications of the aminoglycoside WR279396 on an experimental *Leishmania major*-loaded cutaneous site do not impair the acquisition of immunity. Antimicrob Agents Chemother 54(3):984–990. https://doi.org/10.1128/AAC.01310-09
20. Lecoeur H, Buffet P, Morizot G et al (2007) Optimization of topical therapy for *Leishmania major* localized cutaneous leishmaniasis using a reliable C57BL/6 model. PLoS Negl Trop Dis 1(2):e34. https://doi.org/10.1371/journal.pntd.0000034
21. Sacks DL, Melby PC (2001) Animal models for the analysis of immune response to leishmaniasis. Curr Protoc Immunol . Chapter 19: Unit 19.2. https://doi.org/10.1002/0471142735.im1902s28
22. Sacks DL, Melby PC (2015) Animal models for the analysis of immune response to leishmaniasis. Curr Protoc Immunol 108(19):2): 1–2):24. https://doi.org/10.1002/0471142735.im1902s28
23. Fischer MR, Dominik J, Kautz-Neu K et al (2013) Animal model for cutaneous leishmaniasis. In: Has C, Sitaru C (eds) Molecular dermatology. Methods in molecular biology (methods and protocols), vol 961. Humana Press, Totowa, NJ. https://doi.org/10.1007/978-1-62703-227-8_26
24. Lawyer P, Killick-Kendrick M, Rowland T et al (2017) Laboratory colonization and mass rearing of phlebotomine sand flies (Diptera, Psychodidae). Parasite 24:42. https://doi.org/10.1051/parasite/2017041
25. Baldwin TM, Elso C, Curtis J et al (2003) The site of *Leishmania major* infection determines disease severity and immune responses. Infect Immun 71(12):6830–6834. https://doi.org/10.1128/IAI.71.12.6830-6834.2003
26. Spath GF, Beverley SM (2001) A lipophosphoglycan-independent method for isolation of infective *Leishmania* metacyclic promastigotes by density gradient centrifugation. Exp Parasitol 99(2):97–103
27. Turner PV, Brabb T, Pekow C, Vasbinder MA (2011) Administration of substance to laboratory animals: routes of administration and factors to consider. J Am Assoc Lab Anim Sci 50 (5):600–613
28. Nasseri M, Modabber FZ (1979) Generalized infection and lack of delayed hypersensitivity in BALB/c mice infected with *Leishmania tropica* major. Infect Immun 26(2):611–614

Chapter 8

Multiplex Imaging of Polymicrobial Communities—Murine Models to Study Oral Microbiome Interactions

Jens Kreth, Yasser M. Abdelrahman, and Justin Merritt

Abstract

Similar to other mucosal surfaces of the body, the oral cavity hosts a diverse microbial flora that live in polymicrobial biofilm communities. It is the ecology of these communities that are the primary determinants of oral health (symbiosis) or disease (dysbiosis). As such, both symbiosis and dysbiosis are inherently polymicrobial phenomena. In an effort to facilitate studies of polymicrobial communities within rodent models, we developed a suite of synthetic luciferases suitable for multiplexed in situ analyses of microbial ecology and specific gene expression. Using this approach, it is feasible to noninvasively measure multiple luciferase signals in vivo with both spatial and temporal resolution. In the following chapter, we describe the relevant details and protocols used to establish a biophotonic imaging platform for the study of experimental polymicrobial oral biofilms and abscesses in mice. The protocols described here are specifically tailored for use with oral streptococci, but the general strategies are adaptable for a wide range of polymicrobial infection studies using other species.

Key words Microbiome, Streptococcus, Animal model, Luciferase, Abscess, Biofilm, Dental plaque, Polymicrobial infection, Biophotonic, Bioluminescent imaging

1 Introduction

The oral cavity is a unique host environment that supports host–flora interactions on a variety of mucosal surfaces like the gingiva, hard palate, and tongue as well as on the mineralized nonshedding surfaces of the teeth [1–3]. The oral cavity is also the site of the two most prevalent human dysbiotic diseases, caries (tooth decay) and periodontitis (gum disease).

Symbiosis between the host and microbiota supports oral and systemic health and is achieved through a complex array of microbe–microbe and host–microbe interactions [1, 4, 5]. During a symbiotic relationship with the host, the microbial ecology of the flora favors a high proportion of health-supporting commensal species that prevent the overgrowth of pathobionts via various antagonistic abilities like the production of hydrogen peroxide

Steven Ripp (ed.), *Bioluminescent Imaging: Methods and Protocols*, Methods in Molecular Biology, vol. 2081,
https://doi.org/10.1007/978-1-4939-9940-8_8,

(H_2O_2) [6, 7]. Pathobionts are potentially pathogenic members of the flora community that remain relatively benign provided they are maintained in a low overall abundance [2, 8]. Furthermore, the oral commensal microbiota are comprised of many pioneer colonizer species that are particularly adept at attaching to and inhabiting unoccupied oral surfaces [2, 6]. As such, they play an essential role in preventing niche occupation by less specialized species that are far more likely to trigger pathology to the host.

Oral diseases such as caries and periodontitis are the result of a dysbiotic flora that favors the overgrowth of pathobionts. Like the commensal flora, there are no specific pathobionts known to be singularly responsible for controlling the health or disease status of the host [2, 8]. Thus, both symbiosis and dysbiosis are inherently polymicrobial phenomena controlled by the composition of diverse communities of microorganisms. Modeling the ecological changes associated with polymicrobial diseases is a considerable challenge, as the complex ecology of the flora is itself influenced by a myriad of environmental pressures derived from both host behaviors and host genetics [2, 3]. Therefore, any physiologically relevant experimental model system would need to account for the many complexities of the human oral environment.

Another significant disease associated with the oral cavity is the odontogenic abscess, which similarly exhibits a complex polymicrobial etiology [9, 10]. Dental abscesses develop when oral bacteria breech the integrity of the soft tissues, invading and destroying deeper parts of the oral mucosa. These painful abscesses can cause severe complications requiring surgical intervention and are among the top sources of nontraumatic dental emergencies, especially among children [11, 12]. Currently, there is a limited molecular understanding of odontogenic abscess formation, which is further exacerbated by a dearth of suitable experimental model systems.

In an effort to address some of the major limitations of existing rodent oral disease models, we developed a suite of synthetic luciferases suitable for multiplexing as a biophotonic imaging platform for in situ analyses of oral biofilm ecology [13]. The rodent oral cavity shares many key aspects with the human oral cavity, including innate/adaptive immunity, salivary flow, mineralized teeth, and a resident microbial flora [14, 15]. Thus, it provides a suitable environment to model oral disease development and the resulting clinical aspects of human disease. However, previous animal model systems offer limited options to assess the oral ecology of experimental infections, since the typical PCR- or culture-based approaches used to measure the bacteria are both invasive and often require euthanasia of the host. Consequently, limited information can be gleaned from the spatiotemporal aspects of experimental in vivo oral biofilms. Using multiplexed luciferases for biophotonic imaging, we have demonstrated the feasibility of noninvasively measuring multiple species and specific gene expression in vivo with both spatial and temporal resolution [13]. Our successes with the polymicrobial oral biofilm model subsequently led to an adaptation of this technology for the study of experimental polymicrobial abscesses formed by oral microbes. Our currently employed luciferases have been optimized specifically for expression in oral streptococci and support multiplexing between separate luciferase-expressing species as well as multiplexing within individual organisms for the study of gene expression [13]. To facilitate multiplexed biophotonic imaging, we compared luciferases with easily distinguishable characteristics (i.e., unique enzyme substrates and emission spectra), such as firefly, Cypridina, Renilla, Green Renilla, click beetle green, and Luciola red luciferases (for more information *see* **Note 1**).

Oral streptococci are highly abundant and crucial components of oral biofilms, constituting the vast majority of initial colonizers [3]. A recent metatranscriptomic study determined that >50% of all transcripts detected in supragingival plaques are derived from oral streptococci, which is nearly 5× greater than the next most prevalent genus [16]. Not surprisingly, the ecological significance of oral streptococci for both oral health and disease has been well established by a plethora of clinical and experimental studies [2, 6, 7]. For our model system, we created luciferase reporter strains of the following species: *Streptococcus mutans* (major caries pathobiont), *Streptococcus sanguinis* (pioneer colonizing commensal), *Streptococcus gordonii* (pioneer colonizing commensal), and *Streptococcus anginosus* (periodontitis pathobiont and major component of odontogenic abscesses). In the following chapter, we describe the relevant details and protocols used to establish a biophotonic imaging platform for the study of experimental oral biofilms and abscesses in mice. Furthermore, we have successfully employed these protocols using both of the most commonly employed small animal imaging systems: IVIS (Perkin Elmer) and In-Vivo Xtreme (Bruker). The techniques, reagents, and other materials used in both the oral biofilm and abscess models are grouped when possible to avoid repetition. The following protocols should also be easily adaptable for use in other multiplexed infection models.

2 Materials

2.1 Bacterial Strains

Oral streptococcal bioluminescent reporter strains are constructed in the following wild-type strain backgrounds:

1. *S. mutans* UA159 (genome reference strain) [17].
2. *S. sanguinis* SK36 (genome reference strain) [18].
3. *S. gordonii* DL1 (genome reference strain) [19].
4. *S. anginosus* OUP10 (invasive disease clinical isolate). For more information about *S. anginosus* genomes, *see* [20, 21].

For the polymicrobial abscess model, the following wild-type abscess clinical isolates of *Fusobacterium nucleatum* and *Prevotella nigrescens* are coinfected together with *S. anginosus*:

1. *Fusobacterium nucleatum* ssp. *nucleatum* OHSU59Fn (oral abscess clinical isolate).
2. *Prevotella nigrescens* OHSU79Pn (oral abscess clinical isolate).

2.2 Bacterial Growth Conditions

1. Todd-Hewitt +0.3% (wt/vol) yeast extract (THY) agar plates: 30 g/L TH base (Difco) + 3 g/L yeast extract +15 g/L bacteriological agar. Autoclave at 121 °C for 15 min prior to usage. If antibiotics are supplemented in the agar plates, these should be added after the autoclaved medium has cooled to ≤55 °C.
2. THY liquid medium: Prepare as for #1 above but without the addition of bacteriological agar.
3. Chemically defined medium (CDM): *see* ref. 22, supplement with 0.5% (wt/vol) fructose and 1% (wt/vol) sucrose [13].
4. Erythromycin: 5 μg/ml; used for antibiotic selection of *S. gordonii*, *S. sanguinis*, and *S. anginosus* in liquid cultures and on solid media. For *S. mutans* antibiotic selection, 800 μg/ml kanamycin or 12 μg/ml erythromycin are added.
5. *p*-Chlorophenylalanine (4-CP, Sigma): 0.02 M, For the construction of markerless mutants of *S. mutans* and *S. anginosus*, agar plates are supplemented with 0.02 M 4-CP for counterselection.

2.3 DNA Purification and Manipulation

1. Wizard Genomic DNA purification kit (Promega), DNA Clean and Concentrator 25 kit (Zymo Research), or other similar commercially available DNA extraction kit.
2. Phusion DNA polymerase (New England Biolabs) or Accuprime DNA polymerase (ThermoFisher Scientific) for polymerase chain reaction (PCR) and overlap extension PCR (OE-PCR).
3. PCR and OE-PCR reactions can be performed in any standard thermal cycler. In the described examples, the reactions are performed using a 2720 GeneAmp thermal cycler (Applied Biosystems).

2.4 Luciferase Assays

1. D-Luciferin: 2 mM stock solutions of D-luciferin are created by diluting in 0.1 M citrate buffer (0.04 M sodium citrate dihydrate +0.06 M citric acid dissolved in ddH_2O) adjusted to pH 6.0. Use as substrate for Luciola red luciferase (as well as other beetle luciferases).
2. Coelenterazine or its synthetic analog coelenterazine-h-(NanoLight Technologies): Stock solutions are created by

diluting to 0.75 mg/ml in 100% ethanol. Both are suitable substrates for Green Renilla luciferase (*see* **Notes 2** and **3**).

3. Black or white 96-well microplates (*see* **Note 4**).
4. Microplate reader: in vitro bioluminescence measurements can be obtained in any plate reader capable of measuring luminescence. However, the greatest sensitivity is typically achieved with a dedicated luminometer. In the described examples, samples are measured in a GloMax Discoverer (Promega).

2.5 Animal Housing and Handling

1. BALB/cByJ or C57BL/6J mice (The Jackson Laboratory): House in a maximum of 5 mice/cage. Line cages with standard bedding and provide a source of rodent enrichment.
2. Laboratory Rodent Diet 5001.
3. Isoflurane: 3% (vol/vol), for anesthetization of mice, when required.

2.6 Oral Biofilm Model

1. Weanling 3-week-old BALB/cByJ mice (*see* **Notes 5** and **6**).
2. Mouse drinking water supplements: Depending upon the stage and/or type of experiment, mouse drinking water is supplemented with either 100 μg/ml ampicillin, 5% (wt/vol) sucrose + 5% (wt/vol) fructose, or 5% (wt/vol) sucrose + 5% (wt/vol) fructose + 10% (wt/vol) xylose.
3. Glucose: 20% (wt/vol).
4. Phosphate buffered saline (PBS): 8 g/L NaCl, 0.2 g/L KCl, 1.44 g/L Na_2HPO_4, 0.24 g/L KH_2PO_4, pH 7.4.
5. H_2O_2: 4% (vol/vol).
6. Cotton swabs: Prior to inoculation, mice teeth are cleaned with 2 mm head-width cotton swabs soaked in H_2O_2 (*see* **Note 5**).
7. Small animal imaging system: IVIS (Perkin Elmer), In-Vivo Xtreme (Bruker), or similar instrument.

2.7 Abscess Model

1. 6-Week-old C57BL/6J mice.
2. Depilatory cream: A small patch of fur on the dorsal side of the mice (corresponding to the planned bacteria injection site) is removed using any over-the-counter depilatory cream commonly sold at most drug stores.
3. Insulin syringes: For bacterial injection, Monoject 0.5 ml insulin syringes are used with 28 gauge needles.
4. Modified CDC medium supplemented with laked sheep blood (5%) (*see* ref. 26).
5. Meloxicam: If an analgesic is required, mice are given a daily dose of an oral suspension of meloxicam at 5 mg/kg body weight.

3 Methods

The following section describes the procedures utilized to create synthetic luciferases suitable for multiplexing, the steps required to construct bioluminescent reporter strains of oral streptococci, and the protocols used to measure bacterial growth and gene expression in vivo. Two different polymicrobial infection models are presented, a multispecies oral biofilm model as well as a multispecies abscess model. The abscess model consists of a mixture of *S. anginosus*, *Fusobacterium nucleatum*, and *Prevotella nigrescens*. For the abscess-related methods described here, we will focus specifically upon the *S. anginosus* component of the polymicrobial infection.

3.1 Creation of Synthetic Luciferases

1. We have previously demonstrated the utility of a variety of synthetic codon optimized luciferases for use in oral streptococci, including click beetle green, Luciola red, Green Renilla, and Cypridina luciferases [13]. By combining each of these, it is possible to multiplex at least four distinct bioluminescent signals. For most applications, one or two luciferase signals will suffice. Therefore, we will focus upon our two most commonly utilized luciferases, Luciola red and Green Renilla luciferases (*see* **Note 7**). Since all of these luciferase genes originate from eukaryotic organisms, it is necessary to create codon optimized versions of the luciferase open reading frames (ORFs) to ensure their maximal expression in oral streptococci. The sequences of each luciferase ORF were first codon optimized using the online optimization tool provided by Integrated DNA Technologies (IDT) (www.idtdna.com). *S. mutans* strain UA159 was selected as the reference genome for codon optimization. After the initial optimization, sequences were inspected for the presence of any remaining rare codons and then manually adjusted as needed.
2. Codon optimized luciferase ORFs were synthesized by IDT and then later used as DNA templates for the construction of bioluminescent reporter strains (*see* **Note 8**).

3.2 Construction of Bioluminescent Reporter Strains

1. To yield the greatest sensitivity and accuracy for in vivo biophotonic imaging, luciferase genes should be highly expressed and constitutive. Furthermore, luciferase gene expression should be largely invariant to avoid significant changes in reporter activity due to variable growth conditions. For studies of oral streptococci, we have had the greatest successes with luciferase transcriptional fusions to the following: enolase (*eno*), phosphocarrier protein HPr (*ptsH*), ribosomal protein S21 (*rpsU*), and lactate dehydrogenase (*ldh*).

2. Our constructs are all created using overlap extension PCR (OE-PCR) protocols. For more information on the design and creation of OE-PCR constructs, *see* [23, 24].
3. Two mutagenesis strategies were employed for the construction of constitutive luciferase reporter strains. For *S. mutans* and *S. anginosus*, luciferase ORFs were inserted markerlessly downstream of target genes, whereas for *S. sanguinis* and *S. gordonii*, allelic exchange mutagenesis was employed using antibiotic resistance cassettes. Illustrations of both approaches are shown in Fig. 1. For additional information regarding the use of the IFDC2 cassette for streptococcal markerless mutagenesis, *see* [25].

3.3 Measurement of Bioluminescence in Planktonic and Biofilm Cells

1. Planktonic and biofilm cultures are assayed using a GloMax Discoverer (Promega). Cells are cultured directly within the wells of opaque black or white multiwell plates. Generally, in vitro bioluminescence assays are first conducted to confirm the performance of reporter strains prior to beginning animal imaging studies. It is advisable to test at least five separate clones for each reporter strain constructed. The specific activities for each reporter should be nearly identical. If an outlier is identified among the clones, it should not be used in subsequent studies. Specific activity is determined by normalizing bioluminescence Relative Light Units (RLU) with optical density or colony forming unit (CFU) values (i.e., RLU/OD_{600} or RLU/CFU). Ideally, the specific activity of the reporter strain should also be relatively insensitive to different growth media compositions and culture conditions.
2. *S. mutans* and *S. anginosus* planktonic cells are grown statically at 37 °C to mid-logarithmic phase in an anaerobic chamber (85% N_2, 10% CO_2, and 5% H_2), while *S. sanguinis* and *S. gordonii* are cultured in a CO_2 incubator (5% CO_2).
3. *S. mutans*, *S. sanguinis*, and *S. gordonii* biofilms are grown similarly as described for their respective planktonic cultures, except that cells are diluted 500-fold in medium containing 1% (wt/vol) sucrose prior to 16 h incubation at 37 °C. After the incubation period, the spent overnight medium is aspirated and replaced with an equal volume of prewarmed fresh medium and then further incubated for an additional hour to re-energize cells depleted of ATP and other cellular cofactors (*see* **Note 9**). Afterward, luciferase substrate solutions are added directly into the wells before measuring luciferase activity.
4. For both planktonic and biofilm luciferase assays, stock solutions of luciferase substrates are added in the following ratios: Luciola red, D-luciferin (25 μl per 100 μl culture) and Green Renilla, coelenterazine-h (1.5 μl per 100 μl culture).

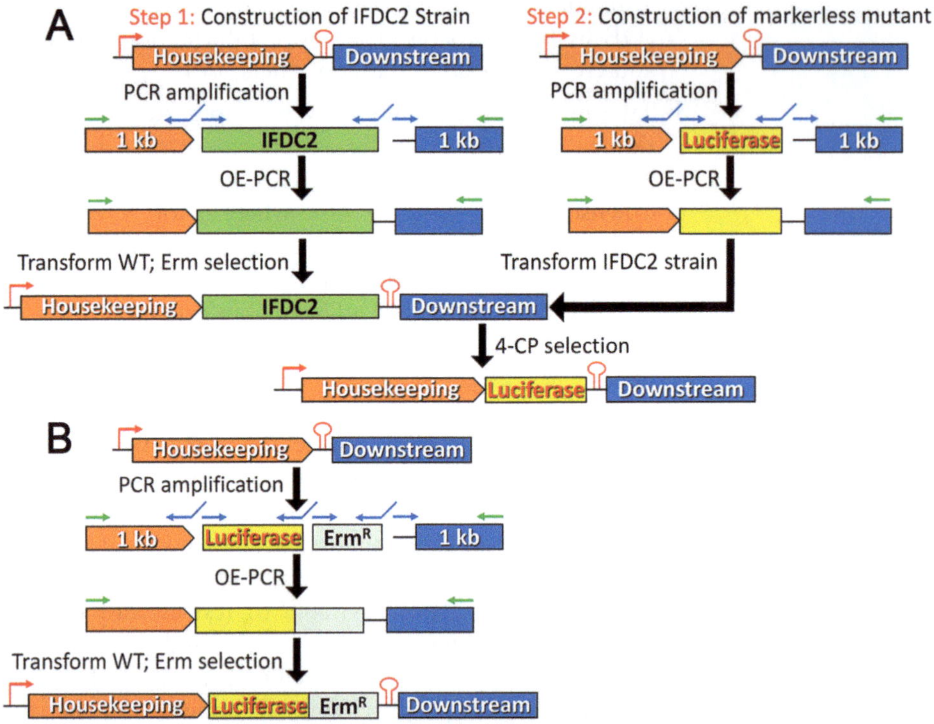

Fig. 1 Construction of markerless and marked luciferase reporter strains. (**a**) Markerless reporter strains are created via a two-step procedure. The first step is to insert the IFDC2 counterselection cassette immediately downstream of the stop codon of a target housekeeping gene (*eno*, *ptsH*, *rpsU*, *ldh*, etc.). The IFDC2 cassette contains markers for both positive selection (erythromycin resistance) and negative selection (4-CP sensitivity). Individual homologous fragments and the IFDC2 cassette are amplified via PCR using primers that contain 5′ reverse complementarity to the adjacent DNA fragments. The three PCR amplicons are then mixed in a single OE-PCR reaction to amplify the assembled construct using the primers shown in green. The resulting OE-PCR amplicon is then transformed into a wild-type background and selected on agar plates supplemented with erythromycin. The second step is to use the same OE-PCR assembly procedure to insert a luciferase ORF immediately downstream of the same target housekeeping gene. This amplicon is transformed into the IFDC2 strain created in step 1 and finally selected on agar plates supplemented with 4-CP. The resulting strain will contain a promoterless luciferase ORF transcriptionally fused to the upstream housekeeping gene in a single polycistronic transcript. (**b**) Marked reporter strains are created by transforming a single OE-PCR product into a wild-type background. Individual homologous fragments, the luciferase ORF, and the erythromycin resistance cassette are amplified via PCR using primers that contain 5′ reverse complementarity to the adjacent DNA fragments. The four PCR amplicons are then mixed in a single OE-PCR reaction to amplify the assembled construct using the primers shown in green. The resulting OE-PCR amplicon is transformed into a wild-type background and selected on agar plates supplemented with erythromycin. The resulting strain will contain a promoterless luciferase ORF and antibiotic cassette transcriptionally fused to the upstream housekeeping gene in a single polycistronic transcript

5. Stock luciferase substrate solutions are added directly into the sample wells, then pipetted vigorously for several seconds to briefly aerate the samples, and finally incubated at room temperature for ~30 s before measuring luciferase activity.

3.4 Animal Handling and Procedures

For instruction on the proper use of the IVIS and In-Vivo Xtreme imagers as well as their system software, please consult the relevant literature provided by the manufacturers. For the oral biofilm and abscess models, optimal image exposure times will vary depending upon the strength of bioluminescence emitted from the reporter bacteria. However, a 3–5 min exposure is usually an effective start point when scouting for the appropriate exposure times.

3.4.1 Oral Biofilm Model

1. For the oral infection of mice using human oral streptococci, the best results are obtained using mice soon after they have been weaned, typically at 21 days (*see* **Note 5**). Prior to experimentation, the mouse drinking water is supplemented with ampicillin (100 μg/ml) for 24 h to suppress the growth of the Gram positive oral flora. The following day, mice are anaesthetized with 3% (vol/vol) isoflurane and then their teeth are manually cleaned using 2 mm head-width cotton swabs soaked in 4% (vol/vol) H_2O_2 (Fig. 2a). Teeth cleaning is performed twice with a brief recovery period given between cleanings. After the second round of cleaning, mice are left in their respective cages for 1 h prior to inoculation with bacteria.
2. To prepare reporter bacteria for inoculation, overnight stationary phase cultures grown in chemically defined medium (CDM) are diluted 1:40 in 10 ml CDM supplemented with 0.5% (wt/vol) fructose (*see* **Note 10**). Cultures are incubated until early logarithmic phase ($OD_{600} \approx 0.3$) before adding 1% (wt/vol) sucrose and then incubating for an additional hr. Afterward, cells are concentrated by centrifugation and resuspended in 1 ml CDM (i.e., concentrated tenfold).
3. To inoculate mice oral cavities, mice are first anesthetized with 3% (vol/vol) isoflurane. Next, an individual mouse is held horizontally with its left side facing down. 30 μl of the concentrated reporter bacteria is pipetted directly into the mouse oral cavity. Immediately afterward, the mouse is returned to the isoflurane anesthesia chamber for an additional 2 min. The mouse should be placed with its inoculation side facing down (i.e., on its left side). The goal is to let the bacterial solution remain in contact with half of the mouth for 2 min. Next, the mouse is returned to its cage for a brief recovery period. During this time, other mice can be inoculated using the same procedure. After all of the mice have been similarly inoculated, the procedure is repeated, except that the mice are inoculated with their right sides facing downward to infect the right sides of their mouths. Once both sides of the mouths have been inoculated, the mice are kept without food or water for 1 h. Afterward, mice are fed a standard Laboratory Rodent Diet 5001 for the remainder of the experiment. At this stage, the mice should also be provided a continuous supply of drinking water supplemented with 5% sucrose and 5% fructose until the experiment has reached its conclusion.

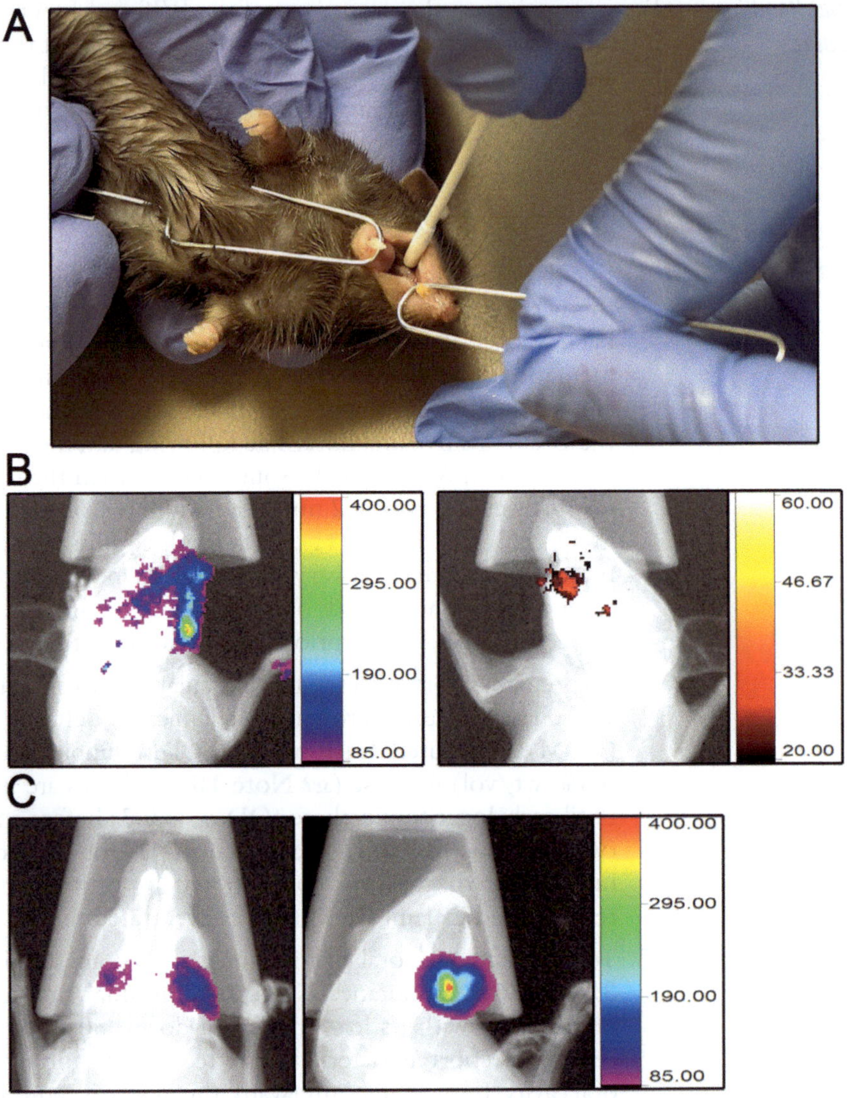

Fig. 2 Infection and imaging for the oral biofilm model. (**a**) To manually clean the mouse teeth, paperclips are used to grasp the mouse incisors and open the mouth. A cotton swab soaked in 4% (vol/vol) H_2O_2 is then used to clean the teeth prior to inoculating with bioluminescent bacteria. The process requires two people. One person holds the mouse and restrains the lower incisors, while the other person restrains the upper incisors and uses the cotton swab to clean the teeth. (**b**) A single mouse was simultaneously infected with bioluminescent reporter strains of both *S. mutans* (left image, Green Renilla luciferase) and *S. gordonii* (right image, Luciola red luciferase). Signal intensity scale bars are located to the right of their respective images. (**c**) A single mouse was orally infected with bioluminescent *S. mutans* and then imaged from two separate angles to localize the sites of colonization in the mouse oral cavity. In the left image, a top-down view of the mouse indicates that *S. mutans* has colonized both halves of the oral cavity, but most of the reporter bacteria are located in the right half (i.e., stronger signal on the right side). In the right image, the mouse is imaged with its left side facing down. Signal localization suggests that *S. mutans* is primarily located on the bottom half of the oral cavity (i.e., mandibular molars). Bioluminescence images in panels B and C were captured using an In-Vivo Xtreme (Bruker) immediately after pipetting the respective luciferase substrates directly into the mouse oral cavity

4. When the mice are ready for imaging, they are first anesthetized using 3% (vol/vol) isoflurane and then 20% (wt/vol) glucose is pipetted into the mouse oral cavities to re-energize the reporter bacteria. After a 15 min incubation period in their cages, the mice are ready for imaging. To activate bioluminescence from the Luciola red reporter, 25 μl of 2 mM D-luciferin stock solution is pipetted directly into the mouse oral cavity. To activate Green Renilla bioluminescence, 1.5 μl of coelenterazine stock solution is added to 25 μl of phosphate buffered saline (PBS) and then the entire 26.5 μl is pipetted directly into the mouse oral cavity. After luciferase substrate(s) have been administered, the mice should be immediately placed into the animal imaging system and analyzed (Fig. 2b, c).

3.4.2 Abscess Model

1. For the polymicrobial abscess infection model, 6-week-old C57BL/6J mice are injected subcutaneously with a suspension of *S. anginosus*, *F. nucleatum*, and *P. nigrescens*.
2. 24 h before infection, the site of injection is treated with a commercially available chemical depilatory cream to remove a small patch of fur (Fig. 3a). The cream should be spread deeply into the fur so that hair removal is as complete as possible.

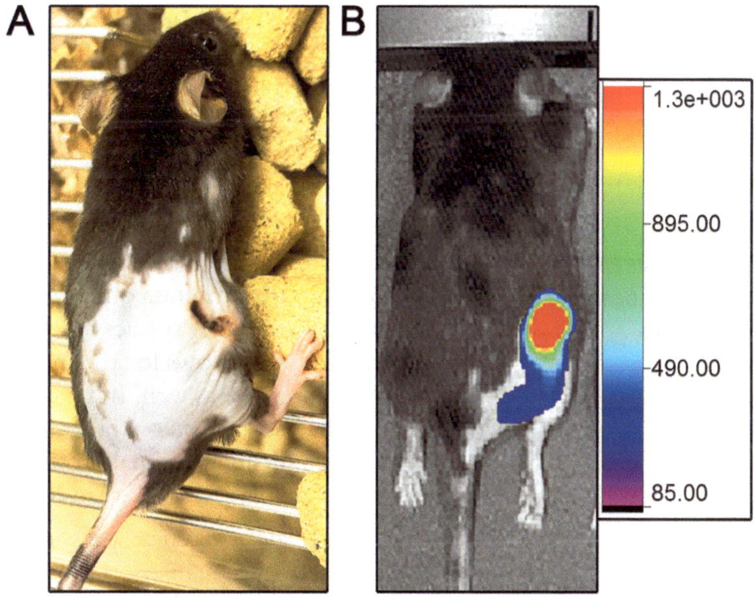

Fig. 3 Infection and imaging for the abscess model. (**a**) 6-day postinjection C57BL/6J mouse previously treated with a depilatory cream. (**b**) IVIS image of the same C57BL/6J mouse 6 days postinjection with a consortium of *S. anginosus*, *F. nucleatum*, and *P. nigrescens*. In this image, only *S. anginosus* emits bioluminescence due to its expression of the Green Renilla luciferase. Coelenterazine-h solution was injected intraperitoneally immediately before imaging. An IVIS (Perkin Elmer) was used to capture bioluminescence from the abscess

Depilatory cream is generally left on the animal for 3–4 min before removing it using tissue paper or a rubber spatula.

3. The bacterial consortium is grown 24–48 h inside an anaerobic chamber (85% N_2, 10% CO_2 and 5% H_2) prior to infection. Bacteria are grown separately on either THY agar (*S. anginosus*) or Modified CDC medium with 5% laked sheep blood [26] (*F. nucleatum* and *P. nigrescens*). Isolated colonies of all 3 species are scraped directly from the agar plates, resuspended in PBS, adjusted to an optical density $OD_{600} \approx 1.0$, and then mixed in a 1:1:1 ratio.
4. Animal injection (*see* **Note 11**): each syringe is resuspended by vigorous shaking to disperse bacterial coaggregates before injection. The injection dose is 0.2 ml. Subcutaneous injections tend to cause superficial abscesses, while intramuscular injections trigger deeper abscesses that are less noticeable from the skin surface. However, intramuscular injections also tend to yield abscesses containing a greater quantity of pus compared to subcutaneous injections.
5. Animal imaging: the coelenterazine-h stock solution is diluted 1:10 in PBS and then delivered via intraperitoneal (IP) injection at a final substrate concentration of 4 mg/kg body weight. Total ethanol in the injection from the original coelenterazine stock solution should not exceed 10% (vol/vol). After luciferase substrate(s) have been administered, the mice should be immediately placed into the animal imaging system and analyzed (Fig. 3b).
6. Following euthanasia, abscess pus is assayed to confirm the organisms present within abscesses (*see* **Note 12**).

3.5 Measurement of Specific Gene Expression

1. Multiplexing luciferase signals provides a straightforward mechanism to simultaneously measure the growth and/or decline of multiple species within experimental infections. However, it also allows for the in situ measurement of changes in gene expression as well. In this case, it is necessary to create a dual luciferase reporter strain. Theoretically, any two easily discernable luciferase signals could be employed. We typically combine Luciola red and Green Renilla luciferases (*see* **Note** 7).
2. To measure specific gene expression, it is necessary to transcriptionally fuse one luciferase ORF to an internal housekeeping control and the other to the target gene of interest. Housekeeping control fusions can be made to any highly expressed constitutive genes, such as those previously described (*eno*, *ptsH*, *rpsU*, *ldh*, etc.). Typically, we fuse these to Luciola red. Since its signal is weaker than that of Green Renilla, the high expression of the housekeeping control gene compensates to yield a robust signal. Conversely, the strong signal output of

Green Renilla luciferase simplifies the detection of target genes with much lower expression relative to the housekeeping gene controls.

3. To calculate specific gene expression, RLU values measured for the gene of interest are normalized using the RLU values measured for the internal housekeeping control (i.e., $RLU_{Target\ Gene}/RLU_{Housekeeping\ Control}$). This approach accounts for the inevitable day to day changes in bacterial population size that occur during experimental infections. Consequently, specific gene expression can be compared between different days during the assay period or even between disparate experimental conditions that impact the growth or viability of the infecting bacteria.

3.6 Exogenous Control of Bacterial Gene Expression During Experimental Infections

1. In many cases, such as virulence gene studies, it can be highly beneficial to exogenously modulate bacterial gene expression within experimental infections. One of the inevitable challenges is determining whether the desired target gene expression pattern is actually occurring in vivo. Using biophotonic imaging, it is possible to make such a determination in situ with both spatial and temporal resolution. In this case, it is again necessary to employ a dual luciferase strain. However, the target gene of interest should be placed under the transcriptional control of an inducible promoter. We previously developed a xylose-inducible gene regulation cassette referred to as Xyl-S1 [27] and recently demonstrated its utility for the exogenous control of gene expression in vivo [13]. Since xylose is nontoxic to mice [28, 29] and is not metabolized by oral streptococci (and many other bacteria) [13], it is a convenient chemical to use in rodent models.
2. For the oral biofilm model, we were able to trigger target gene expression simply by adding 10% (wt/vol) xylose to the mouse drinking water. Furthermore, it was possible to demonstrate using biophotonic imaging that target gene expression could either be induced or repressed depending upon the presence of xylose in the drinking water [13].
3. The calculation of specific gene expression is performed as described in Subheading 3.5.

4 Notes

1. Table 1 provides a comparison of luciferases (click beetle green, Cypridina, firefly, Green Renilla, Luciola red, and renilla) that have been assayed in oral streptococci.

Table 1
Comparison of luciferases assayed in oral streptococci

Luciferase	Substrate	Emission (nm)	ORF (bp)	ATP	Source organism	Comments
Click beetle green[a]	D-Luciferin	537	1629	Yes	*Pyrophorus plagiophthalamus*	Less bright compared to Luciola red
Cypridina[a]	Vargulin	460	1611	No	*Vargula hilgendorfii*	Very bright, currently suitable for in vitro use only
Firefly	D-Luciferin	557	1653	Yes	*Photinus pyralis*	Yellow-green color is not suitable for multiplexing
Green Renilla[a]	Coelenterazine	527	945	No	*Renilla reniformis*	Very bright, suitable for multiplexing
Luciola red[a]	D-Luciferin	609	1647	Yes	*Luciola italica*	Suitable for multiplexing
Renilla	Coelenterazine	420	936	No	*Renilla reniformis*	Very bright, poorer tissue penetration than Green Renilla

[a]NCBI GenBank Accession numbers: Click Beetle Green (MK215076), Cypridina (MK215073), Green Renilla (MK215074), and Luciola Red (MK215075)

2. Both coelenterazine and coelenterazine-h are suitable substrates for Green Renilla reporter strains. However, we have observed up to tenfold higher bioluminescence values with coelenterazine-h. Therefore, it is typically employed as the preferred substrate.
3. We have demonstrated exceptionally strong bioluminescence from Cypridina luciferase reporter strains of oral streptococci. In fact, its signal output is quite similar to that of Green Renilla [13]. Unfortunately, its substrate vargulin is also apparently not freely diffusible across the bacterial cell membrane when in aqueous solutions, unlike both D-luciferin and coelenterazine. It was possible to circumvent this limitation by resuspending reporter bacteria in 100% dimethyl sulfoxide (DMSO, Sigma) immediately prior to vargulin addition. However, this requirement also precludes the use of Cypridina luciferase for in vivo studies, due to the potential toxicity of DMSO. To measure Cypridina luciferase activity in vitro, we centrifuge 100 μl of culture, resuspend the bacteria in 100% (vol/vol) DMSO, and then add 10 μl of vargulin stock solution (1.4 mg ml^{-1} diluted in ethanol) [13]. Vargulin was purchased from NanoLight Technologies.
4. For in vitro bioluminescence measurements, either black or white 96-well microplates can be used. Signal intensities for

white plates can be up to tenfold greater than the same samples assayed in black plates. Therefore, luciferase data should be generated using only black or white plates, as luminescence values generated using different colored plates will not be directly comparable. While 96-well plate color can impact the absolute values of luminescence readings, it does not impact the ratios (i.e., fold-differences) observed between samples, provided all samples are assayed in plates of the same color. Clear 96-well plates are not recommended for bioluminescence measurements because light will leak into adjacent wells and greatly reduce the accuracy of RLU measurements.

5. Successful inoculation of the murine oral cavity with nonresident bioluminescent bacteria is greatly enhanced by reducing competition from the resident mouse oral flora prior to infection. The following are key components of our infection protocol. (1) We infect 3-week-old weanling mice, since weanlings are likely to have a poorly developed natural oral flora. (2) Twenty-four hours before inoculation, the mouse drinking water is supplemented with 100 μg/ml ampicillin to reduce the abundance of live bacteria in the oral cavity, especially gram-positive organisms. (3) We employ a mouse teeth cleaning protocol to remove any preexisting oral biofilm and to reduce the resident oral flora. Cotton swabs soaked in 4% (vol/vol) H_2O_2 proved to be the most effective for this task. H_2O_2 was freshly diluted from a 30% stock solution immediately prior to use. Preferably, two people should be present during teeth cleaning procedures. One person can restrain the mice, while the other cleans the teeth. We found it particularly useful for both people to each use one free hand to place the round end of a paperclip over the top and bottom incisors to hold open the mouse mouth during the teeth cleaning (Fig. 2a).

6. Saliva flow is an important factor to consider for the stable establishment of nonresident foreign bacteria, as low saliva flow is likely to improve the chances of successful colonization. Therefore, we previously employed pilocarpine assays to measure the salivary output of multiple mouse strains to identify the preferred host for our oral biofilm model. Surprisingly, we found the commonly used BALB/CyJ strain to be among the best options, as its salivary output was even lower than that of the murine Sjögren's disease model strain NOD.B10.H2^b [30]. Salivary flow was assayed by injecting 0.2 mg isoproterenol and 0.05 mg pilocarpine per 100 g mouse weight into the peritoneal cavity to induce hypersalivation. Saliva was collected for 3 min immediately after the first signs of salivation appeared postinjection (typically 1–2 min after injection).

7. Wavelengths in the red and far-red region of the spectrum have the greatest ability to penetrate mammalian tissues

[31]. Therefore, when multiplexing in vivo, we prefer to combine Luciola red with Green Renilla luciferase. Green Renilla yields much stronger bioluminescence compared to Luciola red [13], which compensates for the greater tissue absorption of green light. Conversely, Luciola red yields less absolute bioluminescence relative to Green Renilla [13], but its inherent red color exhibits far less absorption by pigments within mouse tissues (i.e., a greater percentage of the signal is available for detection). It is also noteworthy that the bioluminescence emitted by these two enzymes can be distinguished by both substrate choice and signal wavelength.

8. DNA sequences of the codon optimized Luciola red and Green Renilla luciferase ORFs are as follows:

Luciola red ATGGAAACAGAACGTGAA
GAAAATGTTGTCTATGGTCCATTACCATTTTATCC
TATTGAAGAAGGTAGTGCAGGAATCCAGCTTCATAAG
TATATGCAACAGTATGCTAAGCTTGGTGC
TATTGCTTTTAGTAATGCTCTTACAGGTGTTGA
TATTTCTTATCAGCAATATTTTGATAT
TACTTGTCGTTTGGCTGAAGCTATGAAAAATTATG
GAATGAAACCTGAAGGACATATTGCTTTGTGCAGT
GAAAATTGTGAAGAATTTTTTATTC
CAGTTTTGGCTGGTCTTTATATTGGAGT
TACTGTTGCTCCTACAAACGAGATTTATACATTGAGA
GAATTAAATCATTCTTTAGGTATTGCTCAACCAAC
TATTGTATTTAGTTCACGTAAAGGTCTTCCAAAAGTTT
TAGAGGTTCAAAAGACAGTTACATGTATCAAAAC
TATCGTTATTTTAGACTCTAAAGTTAATTTTGGT

GGTTACGATTGTGTTGAAACATTTATTAAAAAG
CATGTTGAATTGGGTTTCCCAGCTACTAGCTTTGTTC
CAATTGATGTTAAAGATCGTAAACATCA
CATTGCTTTGCTTATGAATAGTTCTGGCAGTA
CAGGTTTACCTAAGGGTGTAGAAATTACTCATGAAG
CACTTGTTACACGTTTTTCACATGCTAAAGATC
CAATTTATGGTAATCAAGTTGCACCTGGTACAGC
TATTCTTACAGTGGTTCCTTTTCATCACGGTTTTGG
TATGTTTACTACTTTAGGTTATTTTGCTTGTGGT
TATCGTATCGTTATGTTAACTAAATTTGATGAA
GAGTTGTTTTTGCGTACATTACAGGATTATAAATGTAC
CACTGTTATTTTAGTTCCAACACTTTTTGCTATTCT
TAATCGTAGTGAATTACTTGATAAGTTTGACTTGAG
CAATCTTACAGAAATTGCTTCTGGTGGTGCACCT

CTTGCAAAAGAAATTGGAGAAG
CAGTTGCTCGTCGTTTTAATTTGCCTGGAGTTCGT
CAAGGTTATGGTTTAACTGAAACTACATCTGCTTTTAT
TATTACTCCTAAAGGAGACGATAAGCCTGGTGCTT
CAGGTAAAGTTGTTCCTCTTTTTAAGGTTAAAATTATT

GATCTTGACACAAAAAAGACTTTAGGAGT
TAATCGTCGTGGTGAAATCTGTGTTAAAGGACC
TAGTCTTATGCTTGGTTATACTAACAATCCAGAAGC
TACGCGTGAAACTATTGATGAGGAAGGTTGGCTTCA
TACAGGTGATATTGGATATTATGACGAAGATGAA
CATTTTTTTATCGTTGATCGTTTGAAATCTTTGAT
TAAATATAAAGGATATCAGGTTCCACCTGCTGAATTA
GAAAGTGTATTATTGCAACATCCTAATATTCGTGATG
CAGGTGTTGCAGGTGTTCCAGATTCTGAAGCTGGA
GAGTTACCTGGAGCTGTTGTAGTCATGGAAAAGGG
CAAGACTATGACTGAAAAAGAAATTGTTGATTATGT
TAATAGTCAAGTCGTTAATCATAAACGTTTGC
GTGGTGGAGTTCGTTTTGTAGATGAAGTTCC
TAAGGGTTTAACAGGAAAAATTGATGCTAAAGT
TATCCGTGAAATTTTGAAAAAACCTCAAGCTGGTGGA
TAA

Green Renilla

ATGGCTAGTAAAGTTTATGATCCTGAACAACG
T A A A C G T A T G A T T A C A G G T C C A
CAATGGTGGGCTCGTTGTAAGCAAATGAATGTTTTG
GATAGTTTTATTAATTATTATGATAGTGAAAAACATG
CAGAAAATGCTGTTATTTTCCTTCATGGTAATGCTAC
TAGTAGTTACTTATGGCGTCATGTTGTTCCACACATT
GAACCAGTTGCTCGTTGTATTATTCCTGATCT
TATTGGTATGGGTAAAAGTGGCAAAAGTG
GAAATGGCTCTTATCGTTTATTGGATCATTATAAG
TATTTAACTGCTTGGTTTGAATTGTTAAATCTTC
CAAAAAAAATTATTTTTGTTGGTCATGATTGGGG
TAGTGCTTTAGCTTTCCATTATGCTTATGAACATCAA
GATCGTATTAAAGCTATTGTTCATATG
GAATCTGTTGTTGATGTTATTGAATCTTG
GATGGGTTGGCCAGATATTGAA

GAAGAATTGGCTTTAATTAAATCAGAAGAAGGA
GAAAAAATGGTTCTTGAAAATAACTTTTTTGTTGAAA
CAGTTTTGCCATCTAAAATTATGCGTAAATTGGAAC
CAGAAGAATTTGCTGCTTATTTGGAACCATTTAAG
GAAAAAGGTGAAGTTCGTCGTCCAACTTTGT
CATGGCCTCGTGAAATTCCTTTAGTTAAAGGTG
GAAAGCCTGATGTTGTTGCTATTGTTCGTAATTA
TAATGCTTATTTGCGTGCTAGCGATGATCTTC
CAAAGTTATTCATTGAAAGCGATCCTGGATTCTTTT
CAAATGCTATTGTTGAAGGTGCTAAAAAATTTCCAAA
TACGGAATTTGTTAAAGTTAAAGGTCTTCATTTTCTT
CAAGAAGATGCTCCTGATGAAATGGGTAAGTATAT
TAAATCATTTGTTGAACGTGTTTTGAAGAATGAG
CAACGTTCTATCTAA

9. Most beetle luciferases use D-luciferin as a substrate according to the following reaction:

$$D - \text{luciferin} + O_2 + ATP \rightarrow \text{Oxyluciferin} + CO_2 + H_2O + AMP + \text{light}$$

Due to the requirement for ATP in this reaction, the physiological state of beetle luciferase-expressing reporter bacteria can influence the output of bioluminescence if ATP is limiting in the cell. Renilla luciferase uses coelenterazine as a substrate according to the following reaction:

$$\text{Coelenterazine} + O_2 \rightarrow \text{Coelenteramide} + CO_2 + \text{light}$$

Since Renilla luciferase does not require ATP as a cofactor, its activity is much less sensitive to the physiological state of the cell. However, both beetle and Renilla luciferases function through an oxidation reaction mechanism. Therefore, strict anaerobic conditions may impact signal output. It is worth noting that we routinely grow reporter bacteria in an anaerobic chamber and can still accurately measure luciferase activity. In this case, samples are given a brief aeration simply by vigorously pipetting the samples several times before measuring their reporter activity. If samples are not cultured in strictly anaerobic conditions, it is unnecessary to aerate the samples prior to measurement, even if the bacteria are grown in biofilms. Thus, while O_2 is required for light production, its required concentration for luciferase activity is apparently quite low. We have not observed O_2 concentration as a limiting factor for in vivo studies of the oral biofilm or abscess models.

10. We have observed enhanced oral colonization of reporter bacteria if 25% (vol/vol) sterile human saliva is added to the CDM growth medium. This step is not essential, but in our experience, it improves the outcome.
11. The abscess bacterial consortium will typically coaggregate shortly after mixing. Thus, repeated resuspension by vigorous shaking is required prior to injection. The best results are obtained when single dose injections are prepared inside an anaerobic chamber to maintain strict anaerobic conditions. Every syringe is labeled and filled with 300 μl of the consortium bacterial suspension. Syringes are prepared immediately before use and stored in double sealed ziplock bags containing anaerobic GasPaks (Becton Dickinson) to maintain anaerobic conditions.
12. As an infection control, abscess pus should be collected immediately following euthanasia and then assayed for both the abundance of *S. anginosus*, *F. nucleatum*, and *P. nigrescens* as well as for contamination with exogenous microorganisms. Pus samples are resuspended in PBS, serially diluted tenfold,

spotted onto duplicate modified CDC medium agar plates, and incubated both aerobically and anaerobically at 37 °C.

Acknowledgments

This work was supported by an NIH/NIDCR grant R35DE0282252 to J.M. and NIH/NIDCR grants R01DE021726 and R56DE021726 to J.K.

References

1. Gao L, Xu T, Huang G, Jiang S, Gu Y, Chen F (2018) Oral microbiomes: more and more importance in oral cavity and whole body. Protein Cell 9(5):488–500. https://doi.org/10.1007/s13238-018-0548-1
2. Lamont RJ, Koo H, Hajishengallis G (2018) The oral microbiota: dynamic communities and host interactions. Nat Rev Microbiol 16 (12):745–759. https://doi.org/10.1038/s41579-018-0089-x
3. Sanz M, Beighton D, Curtis MA, Cury JA, Dige I, Dommisch H, Ellwood R, Giacaman RA, Herrera D, Herzberg MC, Kononen E, Marsh PD, Meyle J, Mira A, Molina A, Mombelli A, Quirynen M, Reynolds EC, Shapira L, Zaura E (2017) Role of microbial biofilms in the maintenance of oral health and in the development of dental caries and periodontal diseases. Consensus report of group 1 of the Joint EFP/ORCA workshop on the boundaries between caries and periodontal disease. J Clin Periodontol 44(Suppl 18):S5–S11. https://doi.org/10.1111/jcpe.12682
4. Sudhakara P, Gupta A, Bhardwaj A, Wilson A (2018) Oral dysbiotic communities and their implications in systemic diseases. Dent J (Basel) 6(2):E10. https://doi.org/10.3390/dj6020010
5. Kane SF (2017) The effects of oral health on systemic health. Gen Dent 65(6):30–34
6. Kreth J, Giacaman RA, Raghavan R, Merritt J (2017) The road less traveled - defining molecular commensalism with *Streptococcus sanguinis*. Mol Oral Microbiol 32(3):181–196. https://doi.org/10.1111/omi.12170
7. Redanz S, Cheng X, Giacaman RA, Pfeifer CS, Merritt J, Kreth J (2018) Live and let die: hydrogen peroxide production by the commensal flora and its role in maintaining a symbiotic microbiome. Mol Oral Microbiol 33 (5):337–352. https://doi.org/10.1111/omi.12231
8. Hajishengallis G, Lamont RJ (2016) Dancing with the stars: how choreographed bacterial interactions dictate nososymbiocity and give rise to keystone pathogens, accessory pathogens, and pathobionts. Trends Microbiol 24 (6):477–489. https://doi.org/10.1016/j.tim.2016.02.010
9. Robertson D, Smith AJ (2009) The microbiology of the acute dental abscess. J Med Microbiol 58(Pt 2):155–162. https://doi.org/10.1099/jmm.0.003517-0
10. Siqueira JF Jr, Rocas IN (2013) Microbiology and treatment of acute apical abscesses. Clin Microbiol Rev 26(2):255–273. https://doi.org/10.1128/CMR.00082-12
11. Graham DB, Webb MD, Seale NS (2000) Pediatric emergency room visits for nontraumatic dental disease. Pediatr Dent 22 (2):134–140
12. Jundt JS, Gutta R (2012) Characteristics and cost impact of severe odontogenic infections. Oral Surg Oral Med Oral Pathol Oral Radiol 114(5):558–566. https://doi.org/10.1016/j.oooo.2011.10.044
13. Merritt J, Senpuku H, Kreth J (2016) Let there be bioluminescence: development of a biophotonic imaging platform for in situ analyses of oral biofilms in animal models. Environ Microbiol 18(1):174–190. https://doi.org/10.1111/1462-2920.12953
14. Abe T, Hajishengallis G (2013) Optimization of the ligature-induced periodontitis model in mice. J Immunol Methods 394(1-2):49–54. https://doi.org/10.1016/j.jim.2013.05.002
15. Bowen WH (2013) Rodent model in caries research. Odontology 101(1):9–14. https://doi.org/10.1007/s10266-012-0091-0
16. Peterson SN, Meissner T, Su AI, Snesrud E, Ong AC, Schork NJ, Bretz WA (2014) Functional expression of dental plaque microbiota. Front Cell Infect Microbiol 4:108. https://doi.org/10.3389/fcimb.2014.00108

17. Ajdic D, McShan WM, McLaughlin RE, Savic G, Chang J, Carson MB, Primeaux C, Tian R, Kenton S, Jia H, Lin S, Qian Y, Li S, Zhu H, Najar F, Lai H, White J, Roe BA, Ferretti JJ (2002) Genome sequence of *Streptococcus mutans* UA159, a cariogenic dental pathogen. Proc Natl Acad Sci U S A 99 (22):14434–14439. https://doi.org/10.1073/pnas.172501299
18. Xu P, Alves JM, Kitten T, Brown A, Chen Z, Ozaki LS, Manque P, Ge X, Serrano MG, Puiu D, Hendricks S, Wang Y, Chaplin MD, Akan D, Paik S, Peterson DL, Macrina FL, Buck GA (2007) Genome of the opportunistic pathogen *Streptococcus sanguinis*. J Bacteriol 189(8):3166–3175. https://doi.org/10.1128/JB.01808-06
19. Vickerman MM, Iobst S, Jesionowski AM, Gill SR (2007) Genome-wide transcriptional changes in *Streptococcus gordonii* in response to competence signaling peptide. J Bacteriol 189(21):7799–7807. https://doi.org/10.1128/JB.01023-07
20. Olson AB, Kent H, Sibley CD, Grinwis ME, Mabon P, Ouellette C, Tyson S, Graham M, Tyler SD, Van Domselaar G, Surette MG, Corbett CR (2013) Phylogenetic relationship and virulence inference of *Streptococcus Anginosus* group: curated annotation and whole-genome comparative analysis support distinct species designation. BMC Genomics 14:895. https://doi.org/10.1186/1471-2164-14-895
21. Rahman M, Nguyen SV, McCullor KA, King CJ, Jorgensen JH, McShan WM (2015) Complete genome sequence of *Streptococcus anginosus* J4211, a clinical isolate. Genome Announc 3(6):e01440-15. https://doi.org/10.1128/genomeA.01440-15
22. Merritt J, Niu G, Okinaga T, Qi F (2009) Autoaggregation response of *Fusobacterium nucleatum*. Appl Environ Microbiol 75 (24):7725–7733. https://doi.org/10.1128/AEM.00916-09
23. Aiyar A, Xiang Y, Leis J (1996) Site-directed mutagenesis using overlap extension PCR. Methods Mol Biol 57:177–191. https://doi.org/10.1385/0-89603-332-5:177
24. Pogulis RJ, Vallejo AN, Pease LR (1996) In vitro recombination and mutagenesis by overlap extension PCR. Methods Mol Biol 57:167–176. https://doi.org/10.1385/0-89603-332-5:167
25. Xie Z, Okinaga T, Qi F, Zhang Z, Merritt J (2011) Cloning-independent and counterselectable markerless mutagenesis system in *Streptococcus mutans*. Appl Environ Microbiol 77(22):8025–8033. https://doi.org/10.1128/AEM.06362-11
26. Yoneda S, Loeser B, Feng J, Dmytryk J, Qi F, Merritt J (2014) Ubiquitous sialometabolism present among oral fusobacteria. PLoS One 9 (6):e99263. https://doi.org/10.1371/journal.pone.0099263
27. Xie Z, Qi F, Merritt J (2013) Development of a tunable wide-range gene induction system useful for the study of streptococcal toxin-antitoxin systems. Appl Environ Microbiol 79 (20):6375–6384. https://doi.org/10.1128/AEM.02320-13
28. Johnson SA, van Tets IG, Nicolson SW (1999) Sugar preferences and xylose metabolism of a mammal pollinator, the Namaqua rock mouse (*Aethomys namaquensis*). Physiol Biochem Zool 72(4):438–444. https://doi.org/10.1086/316685
29. Zimmer JP, Lewis SM, Moyer JL (1993) Comparison of gavage, water bottle, and a high-moisture diet bolus as dosing methods for quantitative D-xylose administration to B6D2F1 (*Mus musculus*) mice. Lab Anim 27 (2):164–170. https://doi.org/10.1258/002367793780810423
30. Robinson CP, Yamachika S, Bounous DI, Brayer J, Jonsson R, Holmdahl R, Peck AB, Humphreys-Beher MG (1998) A novel NOD-derived murine model of primary Sjogren's syndrome. Arthritis Rheum 41 (1):150–156. https://doi.org/10.1002/1529-0131(199801)41:1<150::AID-ART18>3.0.CO;2-T
31. Rice BW, Cable MD, Nelson MB (2001) In vivo imaging of light-emitting probes. J Biomed Opt 6(4):432–440. https://doi.org/10.1117/1.1413210

Chapter 9

Bioluminescence Imaging to Study Mature Biofilm Formation by *Candida* spp. and Antifungal Activity In Vitro and In Vivo

Katrien Van Dyck, Patrick Van Dijck, and Greetje Vande Velde

Abstract

The widespread use of indwelling medical devices has increased the number of device-related infections in hospitalized patients. These infections are often associated with the formation of biofilms on the medical implants that are difficult to treat because of their resistance to the classical antifungal drugs. The most common fungi isolated from catheters and other medical devices are *Candida* species. The *Candida* genus contains multiple species of which *C. albicans* and *C. glabrata* are the two most common pathogenic yeasts in humans. A limited number of animal models is available for investigating host–pathogen interactions and testing novel antifungal drugs in vivo against these species. Fungal load in biofilms in these models is traditionally analyzed postmortem, requiring host sacrifice and enumeration of microorganisms from individual biofilms in order to evaluate the amount of colony forming units and the efficacy of antifungal treatment. Bioluminescence imaging (BLI) made compatible with small animal models for in vivo biofilm formation is a valuable tool to follow biofilm development and its treatment longitudinally. Due to the noninvasive nature of BLI, the imaging procedure can be repeated in the same animal, allowing for follow-up of the biofilm growth in vivo without removing the implanted device or detaching the biofilm from its substrate. Although detecting a quantifiable in vivo BLI signal from biofilms formed on the inside of implanted catheters is challenging, BLI proved to be a practical tool in the study of fungal biofilms. This method describes the use of BLI for in vitro and in vivo follow-up of device-related fungal biofilm formation in mice and rats and antifungal activity testing against both *C. albicans* and *C. glabrata* device-associated biofilms. It can further be applied for efficient in vivo screening for interesting genes of the pathogen and the host involved in biofilm formation.

Key words *Candida albicans*, *Candida glabrata*, Biofilm, Bioluminescence imaging, *Gaussia* luciferase, *Firefly* luciferase, Coelenterazine, Luciferin, In vivo, Noninvasive

1 Introduction

The emergence of fungal infections pose an increasing threat to human health. They have become a challenge in clinical practice due to the increasing antimicrobial resistance and the limited amount of available antifungal drugs [1]. Both yeasts (e.g., *Cryptococcus neoformans*) and molds (e.g., *Aspergillus fumigatus*) can

Steven Ripp (ed.), *Bioluminescent Imaging: Methods and Protocols*, Methods in Molecular Biology, vol. 2081,
https://doi.org/10.1007/978-1-4939-9940-8_9,

cause human fungal infections but the majority is caused by the opportunistic pathogenic yeasts of the *Candida* genus [2, 3]. These infections can be superficial and relatively harmless or systemic and life threatening, mainly in immunocompromised patients [4]. *C. albicans* and *C. glabrata* are respectively the first and second most common cause of invasive *Candida* infections in the USA and Europe [5]. Increasing numbers of fungal diseases caused by *Candida* species and other fungi are related to the increasing number of severely ill or immunocompromised patients and in particular to the growing use of implanted medical devices in patients [2, 6, 7].

Biofilm formation is one of the most important virulence factors of these pathogenic fungi. They can adhere to abiotic surfaces such as catheters and prostheses after which they are able to disseminate and get access to internal organs [8]. *Candida* biofilms and their characteristics have been mostly elucidated in vitro [9]. Although they give valuable information, these models cannot account for numerous host and infection-site variables that are important during the development of biofilm-related infection in humans. Therefore, in vivo central venous catheter (CVC) *C. albicans* biofilm models were introduced in rats [10], rabbits [11], and mice [12]. These models are very useful and relevant, but at the same time technically demanding and of very low throughput. To overcome this, a rat subcutaneous foreign body infection model has been optimized to study *C. albicans* biofilms [13, 14]. Still, a limitation to all these models is that the fungal load is traditionally analyzed post mortem by enumeration of the number of colony forming units attached to the catheter fragments and from individual host organs to determine the extent and kinetics of dissemination during infection. Only one time point per animal can be measured and therefore large groups of animals are needed. These end-point assays do not always provide a full understanding of the dynamic steps in *Candida* biofilm development, nor account for interindividual variation.

Noninvasive imaging has the potential to deliver data over the biologically relevant time scale from individual animals in vivo, obviating the need to sacrifice animals at different stages of biofilm development. In recent years, bioluminescence imaging (BLI) has emerged as a powerful method to analyze infectious diseases in animal models and microbial viability in particular [15]. It has become the method of choice to detect fungal presence in living tissues because of the high signal-to-noise ratio [16]. There are several light production systems available in fungi of which two are discussed here: the *Gaussia princeps* luciferase (*gLUC*) and *firefly* luciferase (*fLUC*). A synthetic *C. albicans* codon-optimized version of the gene for the naturally secreted *gLUC* was designed and tagged to the *C. albicans PGA59* gene, which encodes a glycosylphosphatidylinositol (GPI)-linked cell wall protein [17]. Therefore, the luciferase enzyme will be present extracellularly, which avoids potential problems with intracellular delivery of the substrate coelenterazine (CTZ). This expression system has made BLI of superficial *C. albicans* infections possible [17]. In contrast, the *fLUC*, which uses luciferin as a substrate, is expressed intracellularly because it needs ATP as a cofactor for light emission [18]. Recently, the *fLUC* was adapted to improve its thermostability and maximum emission intensity in the red light range for expression in *Candida* cells [16]. Red-shifted light of *fLUC* has the advantage of being less absorbed by host tissues in contrast to the blue light emitted by *gLUC* that has more limited tissue penetration. BLI of both superficial and systemic infections is possible with *fLUC* [19]. In addition, BLI is used to screen the antifungal activity of existing or potentially new antifungal drugs against device-associated *Candida* biofilms. The use of BLI reduces the high effort and ethical burden often associated with these studies and increases the throughput and efficiency, thereby refining preclinical therapy studies [20, 21].

Here we describe a multitemporal noninvasive imaging assay to follow mature biofilm formation and screen the efficacy of antifungals against biofilms in vitro and under in vivo conditions in mice and rats, avoiding the need to sacrifice the animals, which results in statistical, ethical, economic and experimental advantages. The use of in vivo imaging reduces the time, number of animals, and costs for the analysis of biofilm models and increases the versatility of the current animal models to the use of different transgenic mouse strains. The imaging protocol described here will be readily translatable for use with other in vivo biofilm models such as the CVC model, for the assessment of antifungal therapies and the validation in transgenic models to study microbial and host factors influencing biofilm formation [20–22].

2 Materials

2.1 *Candida* Strains

1. *C. albicans* SC5314 (wild-type), a clinical isolate [23]. The genome of this clinical isolate was successfully sequenced.
2. *C. albicans* SKCA23 strain (named SKCA23-*ACTgLUC*) expressing *gLUC* fused to the endogenous *PGA59* gene under the control of *ACT1* (actin) promoter (this promoter is active in the yeast as well as hyphal stage of fungal growth). To construct this strain, *C. albicans* SC5314 was transformed with the Clp10::*ACT1p-gLUC59* plasmid [17], which was integrated in the *RPS10* locus, in order to constitutively express the codon-optimized *Gaussia princeps* luciferase gene (*gLUC*) at the cell wall.
3. *C. glabrata* ATCC2001 (wild-type), used to generate a histidine auxotrophic mutant. The *HIS1* locus was deleted by transforming the wild-type with the *Bam*HI restricted

pUC19 plasmid containing an upstream and downstream fragment of the *HIS1* locus and a nourseothricin resistance cassette [20].

4. *C. glabrata* ATCC2001 *fLUC*. The histidine auxotrophic *C. glabrata* was used to generate a bioluminescent strain with an optimized synthetic *fLUC* gene adapted to the codon usage of *C. glabrata*. The *Bam*HI restricted pUC19 plasmid, containing the promotor of the enolase gene (pENO1) with the luciferase gene and homologous regions of the HIS1 locus for integration, was used to transform *C. glabrata* with histidine prototrophy as selection marker [20].

2.2 Cell Culture and Media

1. Standard cell culture medium: RPMI-1640, with L-glutamine and without sodium carbonate buffered with MOPS, pH 7.0.
2. Phosphate buffered saline: for 1 l of 10× PBS, dissolve 80 g NaCl, 2 g KCl, 14.4 g Na_2HPO_4, 2.4 g KH_2PO_4 in sterile water.
3. Sterile water for preparation of media.
4. YPD agar plates: 1% yeast extract, 2% Bacto peptone, 15% agar, supplemented with 2% glucose in sterile water.
5. Fetal bovine serum (FBS) for incubation of catheter pieces.
6. Cell culture plates: 96-well and 24-well black (or white, *see* **Note 8**) polystyrene plates.
7. Catheters: polyurethane (*see* **Note 1**) triple-lumen intravenous (i.v.) catheters (2.4 mm diameter, Certofix Trio S730, BBraun, *see* **Note 2**), the triple-lumen part cut into 1 cm long fragments.
8. Incubator at 37 °C for static incubation of YPD plates.

2.3 Animals

Before planning animal experiments, seek approval from your local bioethics committee and follow the respective guidelines for animal handling. Always take care that you carry out all aspects of animal experiments in compliance with national and supranational regulations regarding animal handling and welfare.

1. Mice: female Balb/C mice 8 weeks of age, kept in individually ventilated filter top cages with free access to standard food and water ad libitum (*see* **Note 3**).
2. Rats: Sprague-Dawley female rats of 200 g kept in individually ventilated filter top cages with free access to standard food and water ad libitum.
3. Suppress the immune system of the rats (optional in mice: *see* **Note 4**) 24 h before catheter implant and continue during the entire experiment by adding dexamethasone (0.4 mg/l) to their drinking water.

4. When using immune suppression, also add antibiotics (e.g., ampicillin sodium powder 0.5 g/l) to the drinking water to avoid possible bacterial infections.

2.4 Anesthesia and Surgery

1. Anesthesia for mice: prepare 1 ml of anesthetic cocktail by mixing 75 μl of ketamine (Ketamine1000®, 100 mg/ml) with 100 μl of medetomidine (Domitor®, 1 mg/ml) and 825 μl of sterile saline. Administer intraperitoneally (i.p.) 60–80 μl of anesthetic cocktail per 10 g body weight, resulting in a dose of 45–60 mg/kg ketamine and 0.6–0.8 mg/kg medetomidine. For reversal of anesthesia, dilute 50 μl atipamezole (Antisedan®, 5 mg/ml) in 4.95 ml saline, administer i.p. 100 μl per 10 g body weight as antidote, resulting in a dose of 0.5 mg/kg.
2. Anesthesia for rats: prepare 1 ml of anesthetic cocktail by mixing 600 μl of ketamine with 400 μl of medetomidine. Administer i.p. 100 μl of anesthetic cocktail per 100 g body weight, resulting in a dose of 60 mg/kg ketamine and 0.4 mg/kg medetomidine. For reversal of anesthesia, dilute 1 ml atipamezole in 4 ml saline; administer i.p. 100 μl per 100 g body weight as antidote, resulting in a dose of 1 mg/kg.
3. Xylocaine gel (2%) as local anesthetic for the skin.
4. Terramycin/polymyxin-b ophthalmic ointment to apply on the eyes to prevent them from drying/infecting while the animal is under anesthesia.
5. Electric trimmer/razor or disposable razors for shaving the back of the animals.
6. Clean, sterile surgical tools: fine scissors, tweezers, surgical thread, needle and tools for suturing the skin.
7. Sterile gauze, iodoisopropanol (1%) as a disinfectant for the skin.
8. Heating plate or mat to keep the animals warm while they are under anesthesia.
9. Extra clean cages kept warm (on a warm plate or with a lamp) to keep animals separate until they are fully recovered from anesthesia and surgery.

2.5 BLI

1. BLI camera: We have used IVIS 100 and IVIS Spectrum systems (Perkin Elmer, USA) and Living Image software (version 2.50.1 and 4.5.2 for PC, provided by the manufacturer) for measuring and quantifying luminescence in vitro (*see* **Note 5**) and in vivo (*see* **Note 6**).
2. Coelenterazine solution: native coelenterazine (CTZ) was stored at −80 °C according to the manufacturer's instructions. 5 mg/ml CTZ were dissolved in acidified ethanol according to the manufacturer's instructions. This stock solution can be

stored at −80 °C for the duration of the experiments, typically 6–9 days. For in vitro BLI, 6 μM CTZ working solution was prepared by diluting the stock solution 1:2000 in sterile PBS. For in vivo BLI, we prepared a 1.2 mM working solution by diluting the stock solution 1:10 in sterile PBS (*see* **Note 7**). Prepare 100 μl CTZ working solution per catheter trio. Insulin syringes (0.3 ml) were used for injections. CTZ is light-sensitive, always keep it in the dark (e.g., cover recipients with aluminum foil).

3. D-Luciferin solution: For in vitro and in vivo BLI, D-luciferin sodium salt 99% (BC218, Synchem, Germany) was dissolved in sterile saline (0.9% NaCl) at a concentration of 33.3 mg/ml (stock solution). For in vitro experiments, the D-luciferin stock solution was further diluted in sterile saline to 100 mg/ml. For in vivo BLI, 100 μl of the D-luciferin stock solution is needed per injection site. Insulin syringes (0.3 ml) were used for injections. D-Luciferin is light-sensitive, always keep it in the dark (e.g., cover recipients with aluminum foil).

3 Methods

Follow national regulations for handling microbial pathogens class 2. Perform all handlings with open *C. albicans* suspensions in a vertical laminar airflow cabinet (LAF cabinet) to avoid contamination of cell cultures.

3.1 Bioluminescence Imaging of In Vitro Biofilms

3.1.1 In Vitro Biofilm Formation

1. Grow the *C. albicans* or *C. glabrata* strains you want to include in your experiment (e.g., luminescent strain and wild-type as a control) for 8–12 h (e.g., overnight) on YPD plates at 37 °C. Afterward, take a loop of *Candida* cells; dissolve in 1 ml of PBS and wash twice with 1 ml of PBS. Count the cells (e.g., with a Bürker chamber or spectrophotometer) to prepare *Candida* cell suspensions in RPMI-1640 medium at an appropriate cell density (as mentioned below) for subsequent use in the specific experiment.
2. Inoculate a sterile black or white (*see* **Note 8**) 96-well cell culture plate by adding 100 μl per well of a 1×10^7 cells/ml or 1×10^6 cells/ml respectively for a *C. albicans* or *C. glabrata* cell suspension in RPMI-1640. Allow the *Candida* cells to adhere to the bottom of the wells by incubating the plate for 90 min at 37 °C. Wash the wells twice with PBS and then add 200 μl of fresh RPMI-1640 medium to each well. Incubate the plate again at 37 °C for the duration of the experiment to allow mature biofilm formation on the bottom of the wells (*C. albicans* and *C. glabrata* form mature biofilms within 24–48 h).

3. To grow biofilms inside polyurethane catheters, preincubate 1 cm catheter pieces overnight with FBS at 37 °C. Place each catheter fragment in a clean 24-well plate (one catheter piece per well) and add 1 ml of a *Candida* cell suspension (5×10^4 cells/ml for *C. albicans* or 1×10^6 cells/ml for *C. glabrata*). Make sure that catheters are submerged in the medium and not floating on the top. Incubate the plate for 90 min at 37 °C to allow cell attachment onto the substrate. Wash the catheters twice with 1 ml PBS, transfer them to a clean plate, submerge them in fresh RPMI-1640 medium and incubate at 37 °C for the duration of the experiment.

3.1.2 BLI of In Vitro Biofilm Formation

1. Freshly prepare the substrate solution (*see* materials: CTZ or D-luciferin) for in vitro experiments and initialize the BLI camera (*see* **Note 9**).
2. Take the culture plates containing biofilms (formed on the bottom of 96- or 24-well polystyrene cell culture plates or biofilms formed inside catheter pieces) out of the incubator.
3. Pipette the culture medium off and wash the biofilms twice with 1 ml of sterile PBS (*see* **Note 10**), end by taking out all the PBS from the wells.
4. Insert the plate in the camera, set the appropriate field of view (FOV) and find the right imaging position for the plate by taking photographs. Set the settings for acquisition of the BLI image. Typically, we acquired BLI images with a 5–30 s exposure time at medium binning, depending on the signal.
5. Add swiftly but carefully (pipette under a 45° against the well wall) the substrate working solution to the wells (100 μl/well substrate for 96-well plates, 500 μl/well CTZ or 700 μl/well D-luciferin for 24-well plates), preferably with a multichannel pipette (*see* **Note 11**).
6. Immediately thereafter, insert the plates in the BLI-camera and start acquisition of consecutive frames until maximum signal is reached (*see* **Note 12**). Save the acquisitions that you need.
7. Quantify the BLI signal using the processing software of your BLI camera and report as photon flux per second (p/s) for a given region of interest (ROI) with a fixed size, covering the well or catheter (Fig. 1a, b).

3.1.3 In Vitro Screening of Antifungal Activity

1. After 24 h, when mature biofilms are formed, antifungal treatment can be initiated.
2. Add twofold serial dilutions of the antifungal drugs or vehicle (control) to the wells to cover a concentration range of 0.063 mg/l to 64 mg/l.
3. Incubate the plates further for 24 or 48 h at 37 °C before quantification by BLI.

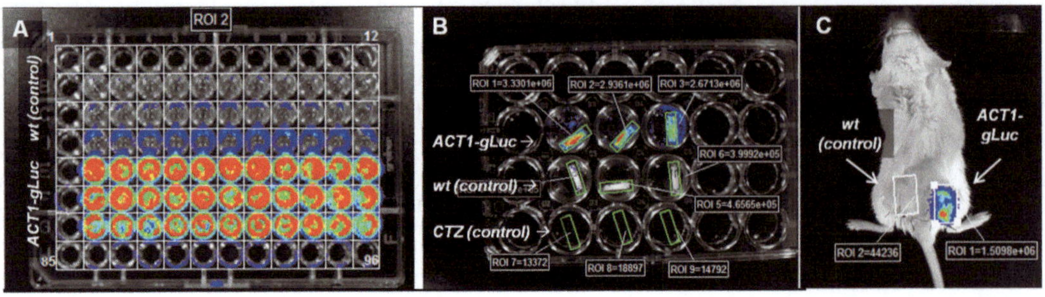

Fig. 1 Regions of interest (ROIs) for BLI signal analysis. (**a**) for the analysis of the BLI signal from biofilms formed on the bottom of wells of a 96-well culture plate, a 8 × 12 grid can be used to quantify the photon flux from each well separately. (**b**) a rectangular ROI is placed over the catheter piece. (**c**) a rectangular ROI is placed over each catheter trio. (Reprinted from [24] with permission from Springer © 2014)

3.2 Bioluminescence Imaging of In Vivo Catheter-Associated Biofilms

3.2.1 In Vivo Biofilm Formation on Implanted Catheters

1. Incubate the catheter pieces in FBS (100%) at 37 °C overnight.
2. Place the serum-coated catheters into separate microcentrifuge tubes and add 1 ml of the *Candida* cell suspension (5×10^4 cells/ml for *C. albicans* and 1×10^6 cells/ml for *C. glabrata* prepared in RPMI-1640).
3. Incubate for 90 min at 37 °C for adhesion of the cells to the catheters.
4. After the adhesion period, place catheters on ice as fast as possible.
5. Wash the catheters twice with 1 ml PBS, transfer them individually into clean, sterile microcentrifuge tubes and place on ice for transport to the surgery room for implantation.
6. Prepare the animal for surgery. Anesthetize the animal by i.p. injection of ketamine–domitor anesthetic cocktail, wait a few minutes until the animal is fully asleep (check by pinching the tail or a paw: the animal should not react to that). Apply ophthalmic ointment on the eyes, shave the lower back of the animal, apply some xylocaine gel on the skin for local anesthesia of the area where you will implant the catheters (wait for a minute to work), disinfect the skin with iodine–isopropanol.
7. Make a small incision in the skin (approximately 0.5–1 cm), carefully dissect the subcutis with a scissor to create two (for mice) or three (for rats) subcutaneous tunnels, make each tunnel about 1.5 cm long and 1 cm wide (Fig. 2).
8. Insert three catheter pieces (inoculated with the same *Candida* strain) in each tunnel, taking care that they fit next to each other in a horizontal arrangement and that they do not lay over each other. This way, you can implant six (for mice) or nine (for rats) catheter fragments in total and test biofilms formed by different *Candida* strains in the same animal (e.g., WT control strain and the corresponding bioluminescent strain).

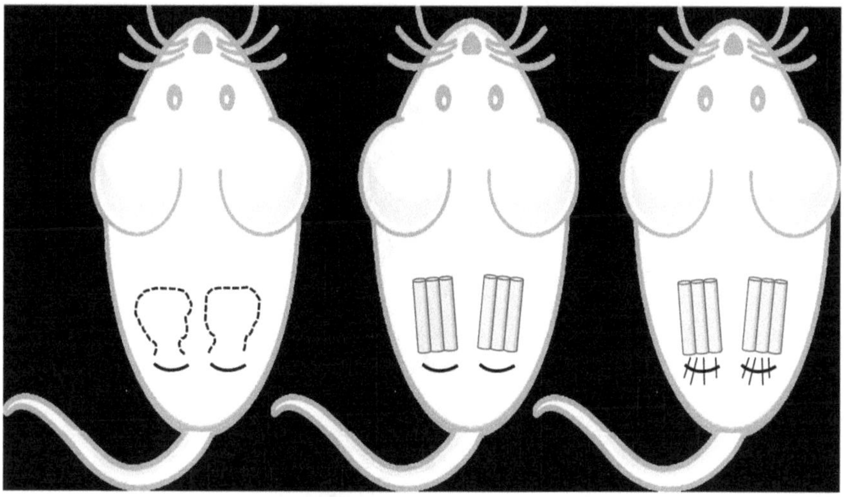

Fig. 2 Catheter implantation. Start by making two incisions in the skin (full black lines in panel **a**). Next, by inserting scissors under the skin through these incisions, dissect the subcutaneous space to create two tunnels of about 1.5 cm long and 1 cm wide (dashed lines in **a**). Make sure that the tunnels do not make contact with each other. Through the incision, insert three catheters next to each other in each subcutaneous space (**b**). Close the incisions in the skin by separate surgical sutures (**c**). Pay attention that you place the sutures well under the catheters and not on top, in order not to block any of the emerging bioluminescence. (Reprinted from [24] with permission from Springer © 2014)

9. Close the incisions by sutures and disinfect the wounds.
10. Wake the animal up by administering antisedan i.p. to reverse anesthesia (*see* **Note 13**).

3.2.2 BLI of In Vivo Biofilm Formation

1. After infection with *C. albicans*, the animals are typically imaged at baseline (day 0, after catheter implantation), on day 2 when a mature biofilm is formed, and on day 6 when a mature biofilm is sustained. For *C. glabrata*, animals are imaged at baseline and continued on day 1, 5 and 7.
2. In vivo BLI: Prepare fresh in vivo substrate working solution (*see* materials: CTZ or D-luciferin) just before the start of the BLI session and initialize the BLI camera (*see* **Note 9**).
3. Anesthetize the animals using an induction box (comes usually with the BLI camera). Use a gas mixture of isoflurane in oxygen, N_2O/O_2 or air, at 2–3% for mice and 3–4% for rats. Rats are best anesthetized and imaged one by one; mice can be put asleep by maximum of four at a time in the induction box and then sequentially imaged.
4. After induction, maintain anesthesia in the induction box and in the imaging chamber at 1.5–2% for mice and 2–3% for rats, at a flow rate of the carrier gas of 0.5 l/min.
5. Before starting the imaging session, put the imaging plate in position A, which corresponds to a FOV of 10 cm. Reassure yourself of the right position of the anesthesia outlets and

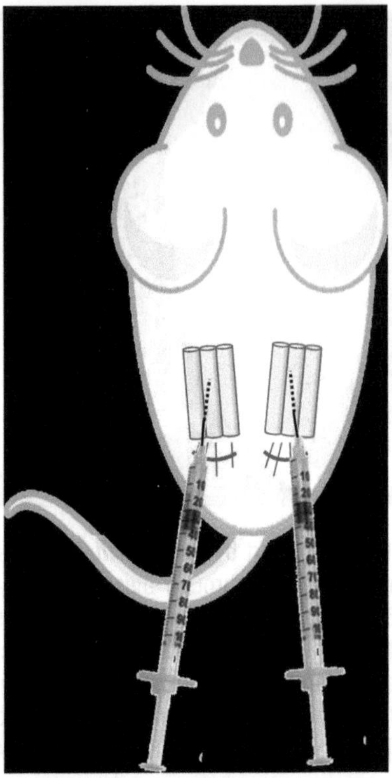

Fig. 3 Topical substrate injection for in vivo BLI. Insert the syringes in such a way that the largest part of the needle is subcutaneously (dashed lines) to avoid leakage of substrate out of the needle entrance point. Apply the substrate on top of the catheter trio (end of dashed line) simultaneously. (Reprinted from [24] with permission from Springer © 2014)

animal by placing a sleeping animal in the box and taking some photographs to confirm the correct imaging position, in the FOV right under the camera.

6. Prepare two syringes containing each 100 μl of the substrate solution. Take one animal and use a nose cone with gas anesthesia to keep it asleep on the bench during the injection. Bring the needles of the syringes one by one in place subcutaneously above the catheters as depicted in Fig. 3 (*see* **Notes 14–16**). Use both hands to inject the substrate simultaneously on top of the catheters (*see* **Note 17**).
7. Immediately after injection, place the animal on the warm plate in the camera box and start the image acquisition. A typical BLI image is depicted in Fig. 4. Acquire consecutive scans with acquisition times ranging from 20 to 60 s (depending on the signal intensity) until the maximum signal intensity is reached (*see* **Note 18**). During the acquisition of the next frame, you can measure the BLI signal of the previously acquired frames by placing a ROI over each catheters trio and measuring the signal intensity (Fig. 1c).

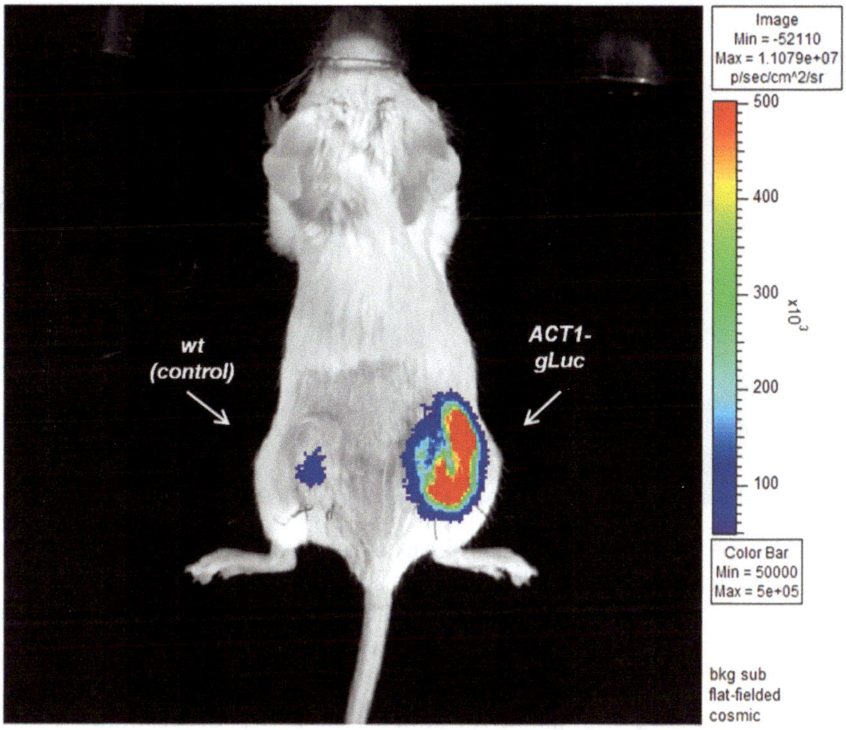

Fig. 4 Typical BLI image from a 2-day-old mature biofilm formed inside 3 catheters implanted left and right under the skin of a mouse. Left: biofilms formed by wild-type *C. albicans* SC5314, right biofilms formed by *C. albicans* SC5314 expressing *gLUC*. (Adapted from [24] with permission from Springer © 2014)

8. Repeat from **step 3** for the next animal(s).
9. Analyze the BLI data and report the BLI signal intensity of each catheter trio as photon flux per second for a rectangular ROI of fixed size placed over each catheter trio.

3.2.3 In Vivo Screening of Antifungal Activity

1. After 2 days of in vivo biofilm formation on the implanted catheters, use BLI to assess the baseline of mature biofilm fungal burden (*see* Subheading 3.2.2).

   Start antifungal treatment and continue the treatment for 7 consecutive days by intravenous or i.p. injection of the antifungal drug or vehicle as a control. A typical BLI image of the result after treatment is depicted in Fig. 5. Appropriate concentrations of the antifungal drug should be determined during the in vitro screening of the antifungal activity.

3.2.4 Explantation of Catheters for Ex Vivo Biofilm Evaluation

After the final imaging time point, recover the catheter pieces from the animals for further ex vivo analysis of the biofilms.

1. Prepare microcentrifuge tubes with 1 ml PBS (each catheter will go into a separate microcentrifuge tube) and label them beforehand.

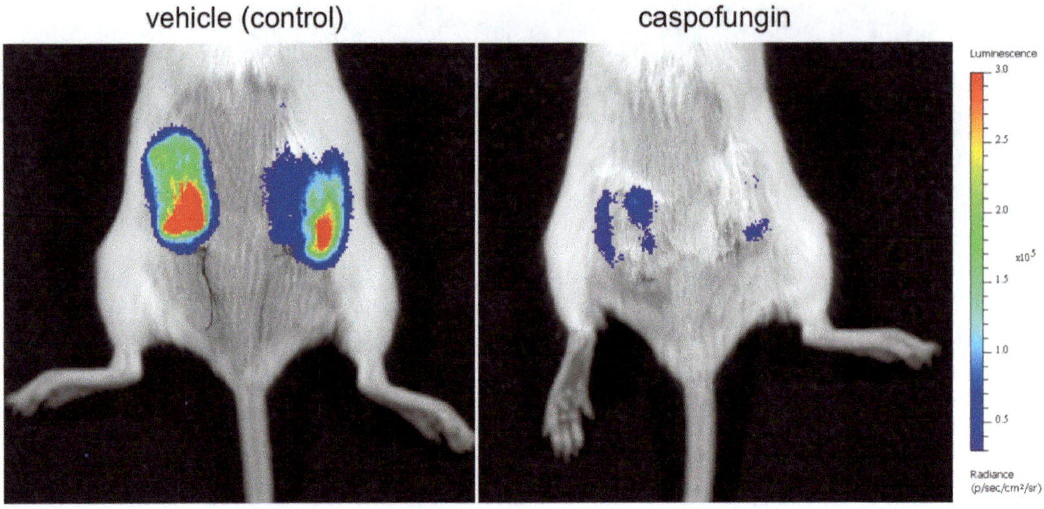

Fig. 5 Typical BLI image from a 9-day old mature biofilm formed inside implanted catheters, and treated with or without caspofungin. In both mice, biofilms were formed by *C. albicans* expressing *gLUC* on 3 catheters implanted left and right. Mice were treated for 1 week with caspofungin or vehicle as a control and imaged thereafter. (Adapted from [21] with permission from Elsevier © 2018)

2. Euthanize the animals by cervical dislocation.
3. Disinfect the skin of the back with 0.5% chlorhexidine in 70% alcohol.
4. Make two new incisions above the catheters and remove the catheter fragments one by one from under the subcutaneous tissue using sterile tweezers (*see* **Note 19**).
5. Wash the catheter twice with 1 ml sterile PBS and put it in the microcentrifuge tube (prepared in **step 1**).
6. Keep the tubes with catheters on ice or in a 4 °C refrigerator for short storage before further processing (i.e., quantification of biomass or with microscopy techniques).
7. Fungal burden quantification: Wash the cell culture plates, in vitro evaluated catheters or explanted catheters twice with 1 ml PBS, sonicate for 10 min at 40,000 Hz in a water bath sonicator (Branson 2210) and vigorously vortex for 30 s. Make 1:10 and 1:100 dilutions and plate 100 μl of the original samples, 1:10 and 1:100 dilutions on YPD agar plates in duplicate. Incubate the plates for 2 days at 37 °C, after which you can count colony forming units (CFUs). Recalculate to fungal load per fragment (CFUs/catheter piece) taking into account the dilution factor and the plated volume.
8. Microscopical evaluation: To examine the three-dimensional structure of the biofilms formed on the inside of the explanted catheters, you can consider performing fluorescence microscopy, confocal laser scanning microscopy or scanning electron microscopy. For these procedures, we refer to [13].

4 Notes

1. Different materials may have different properties for attachment of fungal cells and subsequent biofilm development.
2. Intravenous catheters are often made radiopaque for ease of detection with X-ray imaging, once they are brought into place in the body. The substances added to make catheters radiopaque can elicit phosphorescent properties. We found the majority of catheters commonly used in hospitals to be strongly phosphorescent, which confounds BLI results. Therefore, in our study we used the Certofix catheter from BBraun because it is not phosphorescent. If you would use another type of catheter, carefully check for phosphorescence by imaging it in the BLI camera at regular intervals, and check for an increase in signal after exposing the catheters to light, for example, by opening the scanner door and resuming BLI scanning.
3. We use Balb/C mice as this is a common background strain for many transgenic mouse models. For BLI, it is important that they are white (or nude) as the pigmentation of the skin, even when shaving the animals, absorbs a substantial amount of the BLI signal.
4. The biofilms will also develop without immune suppression of the host, but the fungal load on the catheters may vary substantially in rats. When suppressing the immune system of rats, the fungal load recovered from the catheters is much more consistent [22], reducing the standard deviations and hence reducing the amount of animals needed to reach a given power in your experiment. Therefore, we recommend you to use the protocol for immune suppression before you start the catheter implantation in rats. For mice, we did not observe a similar reduction in variation of CFU counts when modulating the immune suppression [21].
5. For measuring luminescence in vitro, you can also use any luminometer or essentially any plate reader that is suitable for measuring luminescence. Some can even be equipped with injectors for controlled addition of the substrate. The advantage of placing the plates in the BLI system is that a luminometer/plate reader will not provide an image of your plate.
6. The IVIS systems come with a very comprehensive manual, it is worth reading through (the first part) of the manual and you will be able to get started with the camera and the software easily.
7. When diluting the stock solution to make the in vivo CTZ working solution, a small precipitate will form. According to the manufacturer this should not happen, but in our hands

using the prescribed dilution protocol, we could never avoid having some extent of precipitation. As far as we know, other users experienced the same problem. It is imperative to dilute the CTZ immediately prior to injection as this will limit the precipitation.

8. For BLI measurements, it is recommended to use black or white cell culture plates, with or without transparent bottom. When not using these black or white plates, there will be BLI signal scattering from adjacent wells into the well you are measuring in case you have high signal intensities. In case you do not have a black or white plate at hand, you can also use a transparent plate and poor black drawing ink in between wells to reduce this "cross talk." This is not ideal as there will still be some "cross talk" where the wells are touching each other, as it will have some influence on the quantification of your BLI results. After your BLI experiment is finished, you can use a pipette to take out the ink for later reuse.

9. Starting up BLI camera: Before you start imaging, you need to login and initialize the system (button: "initialize"). If your BLI system is only used sporadically, it can be kept off in between imaging sessions. In that case, it will need to cool down before you can start imaging, which can take up to several hours depending on the system. In most cases, however, the system is left on to keep the CCD camera (partially) cooled and to take background data overnight. Initializing the system takes then only few minutes and you will be able to start almost immediately with your imaging experiment.

10. Especially when you are dealing with biofilms formed on the bottom of wells, take care not to detach the biofilm from the bottom of the plates while washing.

11. When doing in vitro BLI measurements, you could consider making fresh substrate and then incubate it in the dark for ~30 min before you add it to the wells. This way, the difference in auto-oxidation of the substrate solution (very relevant in the case of CTZ) will be minimal in case you cannot add the substrate to all the wells simultaneously. When you set up your experiments, always take nonluminescent wild type-controls and control wells where you only add substrate along in your analysis for comparison with unspecific background signal. On the other hand, if you are interested in measuring the signal kinetics, make sure you add the substrate "as simultaneously as you can" (when pipetting with a multichannel pipette, make sure that in every pipetting move you take along appropriate controls) and start image acquisition as fast as you can after substrate addition.

12. In practice, you will almost immediately reach the maximum signal when performing in vitro BLI measurements as diffusion of the substrate into the biofilms is fast, after which the signal will steadily decrease. Depending on your research question, you can take about three to four consecutive frames at maximum signal with the settings of your choice or keep acquiring short frames during a certain period if you are interested in signal kinetics.
13. We do not recommend keeping the animal under ketamine/domitor anesthesia and proceed to BLI immediately after surgery while it is still anesthetized. We recommend letting the animal completely recover from the injection anesthesia before starting isoflurane anesthesia for the imaging session. The latter procedure was well tolerated by the animals.
14. We evaluated signal kinetics in vivo upon different administration routes of the CTZ (intravenous, intraperitoneal, subcutaneous) and compared the BLI signal. We found that CTZ does not have a favorable biodistribution upon i.v. or i.p. injection. For this reason, we recommend to apply CTZ topically. Here, we applied D-luciferin topically in analogy to CTZ administration, but we recommend to always optimize the administration route for your own in vivo model system.
15. Upon substrate injection, it is crucial that no substrate leaks out of the little hole you make in the skin upon inserting the syringe. That is why it is crucial to keep ~1 cm of the needle under the skin when you inject, as shown in Fig. 3. CTZ shows some extent of auto-oxidation. Therefore, any "leaked" CTZ on the skin of the animal would give rise to unspecific BLI signal that would significantly influence the BLI measurement.
16. Injecting the substrate subcutaneously is obvious at the first time-point, as there is a lot of subcutaneous space because of the tunnels created during surgery. However, this will heal over time and there will be fibrous tissue formed around the catheters, making injecting subcutaneously less evident, especially when doing BLI at later time points (6–9 days after catheter implantation). Take care not to inject intracutaneously. In this case, you will notice immediately by a "bobble" appearing due to the intradermal substrate injection. No or only a very low signal will be seen.
17. In case experiments are performed on rats that have three (or more) tunnels with catheter trios, ask a colleague to assist you with injecting the substrate in all the tunnels simultaneously.
18. At the first time-points, the maximum signal is reached in less than 1 min after CTZ administration. For D-luciferin, the maximum signal is reached about 15 min after injection. The

later after catheter implantation surgery, the longer it takes until this maximum signal is reached, presumably because of the increase of fibrous tissue around the catheters that makes the substrate diffuse more slowly toward the biofilms.

19. Make sure to decontaminate the tweezers in between catheter pieces having biofilms from different *Candida* strains.

Acknowledgments

This work was funded by the Fund for Scientific Research Flanders (FWO) Research community on biology and ecology of bacterial and fungal biofilms (FWO: WO.026.11N), FWO research grant G.0691.15N, and FWO PhD fellowship to KVD (FWO 1181818N) and KU Leuven IF grants STG/15/24 and C32/18/010. We thank Christophe d'Enfert for providing us with the Clp10::ACT1p-gLUC59 plasmid. All imaging experiments were conducted at the Molecular Small Animal Imaging Center (MoSAIC) core facility of the KU Leuven. (Parts of this protocol have been updated and reprinted from [24] by permission from Springer © 2014.)

References

1. Brown GD, Denning DW, Levitz SM (2012) Tackling human fungal infections. Science 336 (6082):647. https://doi.org/10.1126/science.1222236
2. Nucci M, Marr KA (2005) Emerging fungal diseases. Clin Infect Dis 41(4):521–526
3. Warnock DW (2006) Fungal diseases an evolving public health challenge. Med Mycol 44 (8):697–705
4. Wisplinghoff H, Bischoff T, Tallent SM, Seifert H, Wenzel RP, Edmond MB (2004) Nosocomial bloodstream infections in US hospitals: analysis of 24,179 cases from a prospective nationwide surveillance study. Clin Infect Dis 39:309–317
5. Periroth J, Choi B, Spellberg B (2007) Nosocomial fungal infections: epidemiology, diagnosis, and treatment. Med Mycol 45 (4):321–346
6. Eggimann P, Garbino J, Pittet D (2003) Epidemiology of *Candida* species infections in critically ill non-immunosuppressed patients. Lancet Infect Dis 3:685–702
7. Eggimann P, Garbino J, Pittet D (2003) Management of *Candida* species infections in critically ill patients. Lancet Infect Dis 3 (12):772–785
8. Costa-Orlandi CB, Sardi JCO, Pitangui NS, de Oliveira HC, Sorzoni L, Galeane MC, Medina-Alarcón KP, Melo WCMA, Marcelino MY, Braz JD, Fusco-Almeida AM, Mendes-Giannini MJS (2017) Fungal biofilms and polymicrobial diseases. J Fungi (Basel) 3(2): E22
9. Nett J, Andes D (2006) *Candida albicans* biofilm development, modeling a host-pathogen interaction. Curr Opin Microbiol 9:340–345
10. Andes D, Nett J, Oschel P, Albrecht R, Marchillo K, Pitula A (2004) Development and characterization of an *in vivo* central venous catheter *Candida albicans* biofilm model. Infect Immun 72:6023–6031
11. Schinabeck MK, Long LA, Hossain MA, Chandra J, Mukherjee PK, Mohamed S, Ghannoum MA (2004) Rabbit model of *Candida albicans* biofilm infection: liposomal amphotericin B antifungal lock therapy. Antimicrob Agents Chemother 48:1727–1732
12. Lazzell AL, Chaturvedi AK, Pierce CG, Prasad D, Uppuluri P, Lopez-Ribot JL (2009) Treatment and prevention of *Candida albicans* biofilms with caspofungin in a novel central venous catheter murine model of candidiasis. J Antimicrob Chemother 64:567–570

13. Řičicová M, Kucharíková S, Tournu H, Hendrix J, Bujdáková H, Van Eldere J, Lagrou K, Van Dijck P (2010) *Candida albicans* biofilm formation in a new in vivo rat model. Microbiol 156:909–919
14. Van Wijngaerden E, Peetermans WE, Vandersmissen J, Van Lierde S, Bobbaers H, Van Eldere J (1999) Foreign body infection: a new rat model for prophylaxis and treatment. J Antimicrob Chemother 44:669–674
15. Hutchens M, Luker GD (2007) Applications of bioluminescence imaging to the study of infectious diseases. Cell Microbiol 9:2315–2322
16. Dorsaz S, Coste AT, Sanglard D (2017) Red-shifted firefly luciferase optimized for *Candida albicans* in vivo bioluminescence imaging. Front Microbiol 8:1478
17. Enjalbert B, Rachini A, Vediyappan G, Pietrella D, Spaccapelo R, Vecchiarelli A, Brown AJ, d'Enfert C (2009) A multifunctional synthetic *Gaussia princeps* luciferase reporter for live imaging of *Candida albicans* infections. Infect Immun 77:4847–4858
18. Doyle TC, Nawotka KA, Kawahara CB, Francis KP, Contag PR (2006) Visualizing fungal infections in living mice using bioluminescent pathogenic *Candida albicans* strains transformed with the firefly luciferase gene. Microb Pathog 40:82–90
19. Papon N, Courdavault V, Lanoue A, Clastre M, Brock M (2014) Illuminating fungal infections with bioluminescence. PLoS Pathog 10(7): e1004179
20. Persyn A, Rogiers O, Brock M, Vande Velde G, Lamkanfi M, Jacobsen ID, Himmelreich U, Lagrou K, Van Dijck P, Kucharíková S (2019) Monitoring of fluconazole and caspofungin activity against in vivo *Candida glabrata* biofilms by bioluminescence imaging. Antimicrob Agents Chemother 63:e01555-18
21. Vande Velde G, Kucharíková S, Van Dijck P, Himmelreich U (2018) Bioluminescence imaging increases in vivo screening efficiency for antifungal activity against device-associated *Candida albicans* biofilms. Int J Antimicrob Agents 52:42–51
22. Kucharíková S, Tournu H, Holtappels M, Van Dijck P, Lagrou K (2010) In vivo efficacy of anidulafungin against *Candida albicans* mature biofilms in a novel rat model of catheter-associated candidiasis. Antimicrob Agents Chemother 54:4474–4478
23. Gillum AM, Tsay EYH, Kirsch DR (1984) Isolation of the *Candida albicans* gene for orotidine-5′-phosphate decarboxylase by complementation of *S. cerevisiae ura3* and *E. coli pyrF* mutations. Mol Gen Genet 198:179–182
24. Vande Velde G, Kucharíková S, Van Dijck P, Himmelreich U (2014) Bioluminescence imaging of fungal biofilm development in live animals. Methods Mol Biol 1098:153–167

Part III

In Vivo Small Animal/Organism Imaging Applications

Chapter 10

Bioluminescence Imaging of Neuroinflammation in a Mouse Model of Parkinson's Disease

Maria Eugenia Bernis and Gültekin Tamgüney

Abstract

In Parkinson's disease (PD) and related disorders pathological alpha-synuclein has been discussed to propagate via a prion-like mechanism in the CNS. The application of exogenous alpha-synuclein fibrils via injection to animal models of PD has been shown to be a useful method to study prion-like propagation of pathological alpha-synuclein and of transmission pathways that play a critical role in recapitulating characteristics of synucleinopathies. Using bigenic mice expressing mutant human alpha-synuclein in neurons and firefly luciferase in astrocytes we showed that transmission via the tongue and the peritoneum represent entrance points for pathological alpha-synuclein to invade the CNS. Here we present a method to quantify astrogliosis by bioluminescence imaging in an animal model of PD. This method allows noninvasive tracking of the neuroinflammatory process that often precedes neurological signs of disease and represents an alternative to behavioral or histological and biochemical analysis to detect disease.

Key words Alpha-synuclein, Bioluminescence imaging, Neuroinflammation, Synucleinopathy, Neuroinvasion, Parkinson's disease, Peripheral inoculation, Prion-like

1 Introduction

Several age-related neurodegenerative disorders are characterized by the deposition of aberrantly folded endogenous proteins. Parkinson's disease (PD), dementia with Lewy bodies (DLB), and multiple system atrophy (MSA) are neurodegenerative disorders that have been pathologically classified as synucleinopathies. Analogous to infectious prions, pathological alpha-synuclein has been shown to spread via a prion-like mechanism. Based on this mechanism toxic self-propagating forms of pathological alpha-synuclein induce the conversion of endogenous, naive alpha-synuclein in interconnected and neighboring areas, thereby propagating the pathology [1]. Compelling evidence indicates that alpha-synuclein can spread from diseased to healthy cells. Indeed, graft-derived dopamine neurons can develop Lewy body (LB) pathology several years after transplantation [2–4]. Moreover, growing evidence has

Steven Ripp (ed.), *Bioluminescent Imaging: Methods and Protocols*, Methods in Molecular Biology, vol. 2081,
https://doi.org/10.1007/978-1-4939-9940-8_10, © Springer Science+Business Media, LLC, part of Springer Nature 2020

demonstrated that direct injection of misfolded alpha-synuclein into the brain or the periphery, for example, the intestinal wall or hind limb muscle, results in a spread of alpha-synuclein pathology to distal parts of the CNS [5–17]. With this method we analyzed the transmission of pathological alpha-synuclein via peripheral routes in more detail and addressed the question whether misfolded alpha-synuclein can neuroinvade the CNS after a single intraglossal or intraperitoneal injection [17], a feature that previously had been shown for prions but not for pathological alpha-synuclein [18]. After injection of prions into the tongue, neuroinvasion of the CNS is achieved via propagation along the hypoglossal nerve of the tongue, leading to the nucleus of the hypoglossal nerve, which is located in the brain stem [18]. As a mouse model we chose Tg(M83$^{+/-}$:*Gfap*-luc$^{+/-}$) mice. These mice overexpress the A53T mutant of human alpha-synuclein under the control of the prion promoter [19]. Additionally, these mice express firefly luciferase under the control of the *Gfap* promoter [20], which allows monitoring of astrogliosis by bioluminescence imaging as we have previously shown in the brain of prion-infected mice [21]. In our hands, bigenic Tg(M83$^{+/-}$:*Gfap*-luc$^{+/-}$) mice did not develop disease until 23 months of age as has been shown by others [10]. Peripheral injection of alpha-synuclein fibrils via the tongue or the peritoneum induced neuropathology in the brain and spinal cord of bigenic Tg(M83$^{+/-}$:*Gfap*-luc$^{+/-}$) mice (Table 1) (Fig. 1). The accumulation of pathological alpha-synuclein in diseased animals was accompanied by neuroinflammatory changes, which were detected by immunofluorescence staining of brain sections for GFAP, a marker of reactive astrocytes or IBA-1 (ionized calcium-binding adapter molecule 1), a marker for microgliosis. Diseased animals showed astrogliosis and microgliosis in the presence of deposits of phosphorylated alpha-synuclein (Fig. 2). Moreover, bioluminescence imaging of all Tg(M83$^{+/-}$:*Gfap*-luc$^{+/-}$) mice intraperitoneally inoculated with alpha-synuclein fibrils showed increased radiance in the brain long before they developed signs of neurologic disease (Fig. 3). Since luciferase expression is driven from the *Gfap* promoter, increased radiance in Tg(M83$^{+/-}$:*Gfap*-luc$^{+/-}$) mice is indicative of astrogliosis. Transmission via the tongue and the peritoneum represent further entrance points for pathological alpha-synuclein to invade the CNS [17]. Quantification of reactive astrocytes by bioluminescence imaging facilitates the tracking of the neuroinflammatory process over time and represents a noninvasive alternative to histological and biochemical analysis.

Table 1
Inoculation experiments

Mouse line	Inoculum (μg)	Inoculation route	No. of mice with disease/no. of mice inoculated	Mean survival time ± SD (day)
Tg(M83$^{+/-}$:*Gfap*-luc$^{+/-}$)	Human α-synuclein fibrils (50)	Intraperitoneal	4/5	229 ± 17
Tg(M83$^{+/-}$:*Gfap*-luc$^{+/-}$)	PBS	Intraperitoneal	0/5	420
Tg(M83$^{+/-}$:*Gfap*-luc$^{+/-}$)	Human α-synuclein fibrils (10)	Intraglossal	1/5	285
Tg(M83$^{+/-}$:*Gfap*-luc$^{+/-}$)	PBS	Intraglossal	0/4	420

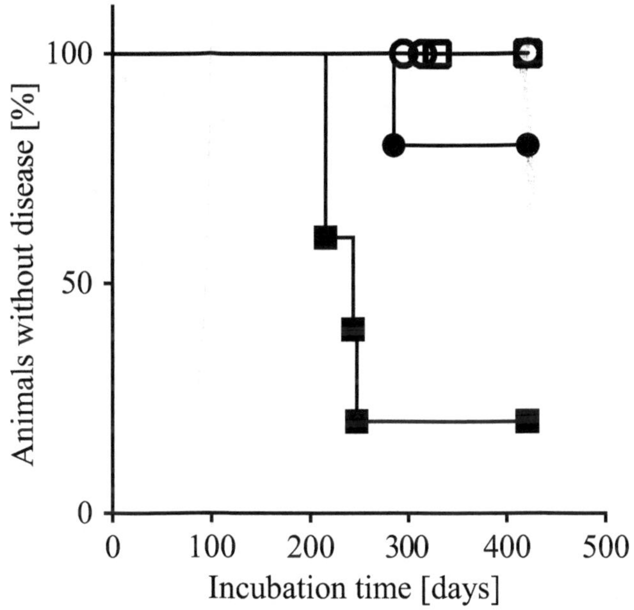

Fig. 1 Kaplan–Meier survival curves for Tg(M83$^{+/-}$:*Gfap*-luc$^{+/-}$) mice that developed signs of neurologic disease after peripheral injection with alpha-synuclein fibrils. Four of five Tg(M83$^{+/-}$:*Gfap*-luc$^{+/-)}$ mice developed signs of neurologic disease in 229 ± 17 days (black squares) after intraperitoneal injection with alpha-synuclein fibrils, whereas none of the PBS-injected control mice developed signs of neurologic dysfunction within 420 days (white framed squares). One mouse of five died after 285 days (black circles) after intraglossal inoculation with alpha-synuclein fibrils. In contrast, none of the PBS-injected control mice developed neurologic disease or spontaneously died (black framed circles)

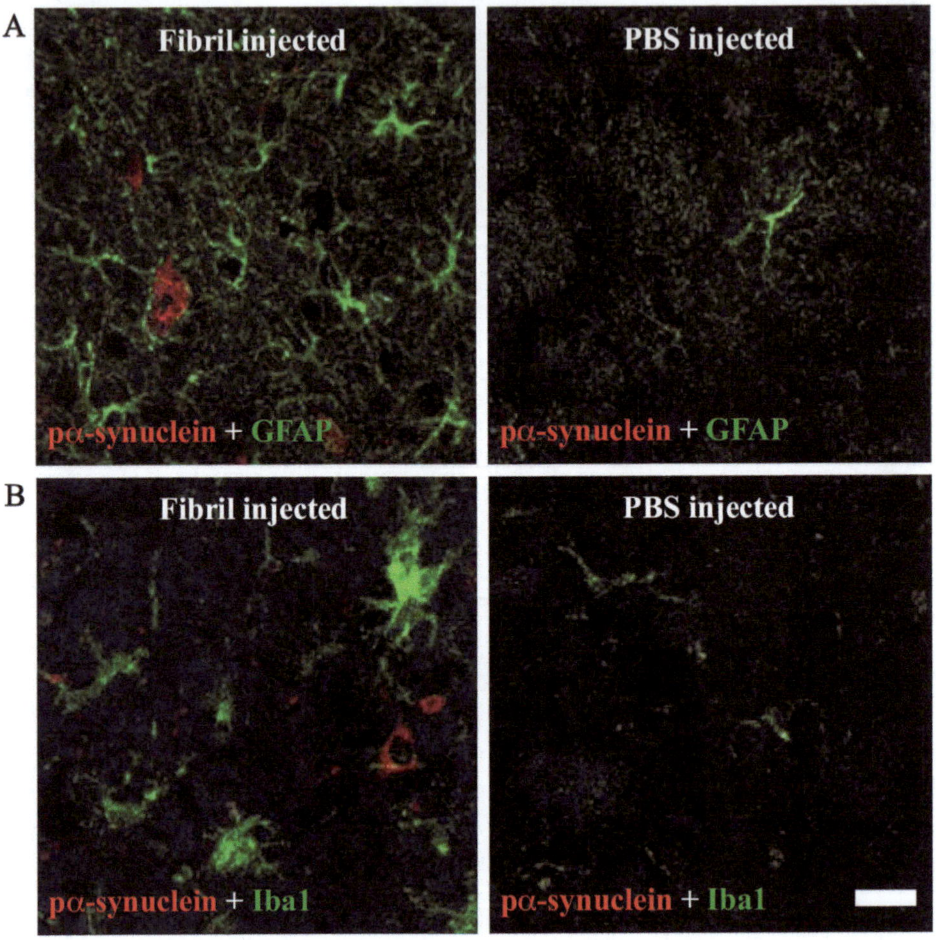

Fig. 2 Peripheral challenge with alpha-synuclein fibrils induced gliosis in the CNS of Tg(M83$^{+/-}$:*Gfap*-luc$^{+/-}$) mice. (**a**) Immunofluorescence analysis of brain tissue sections of sick animals with an antibody against GFAP, which is a marker for astrocytes, showed astrocytic gliosis in areas with deposits of phosphorylated alpha-synuclein, which were detected here in the brain stem with the pSyn#64 antibody. PBS-injected control mice did not harbor any activated astrocytes in their CNS. (**b**) Staining with an antibody to IBA-1, which is a marker for microglia, demonstrated microgliosis in areas with deposits of phosphorylated alpha-synuclein in sick mice. PBS-injected healthy, control mice did not show signs of microgliosis in their CNS

2 Materials

Prepare all solutions using ultrapure water (deionized water purified to attain a resistivity of 18 MΩ cm at 25 °C) and analytical grade reagents. Prepare and store all reagents at room temperature (unless indicated otherwise). Diligently follow all institutional disposal regulations when discarding waste materials. We do not add sodium azide to reagents.

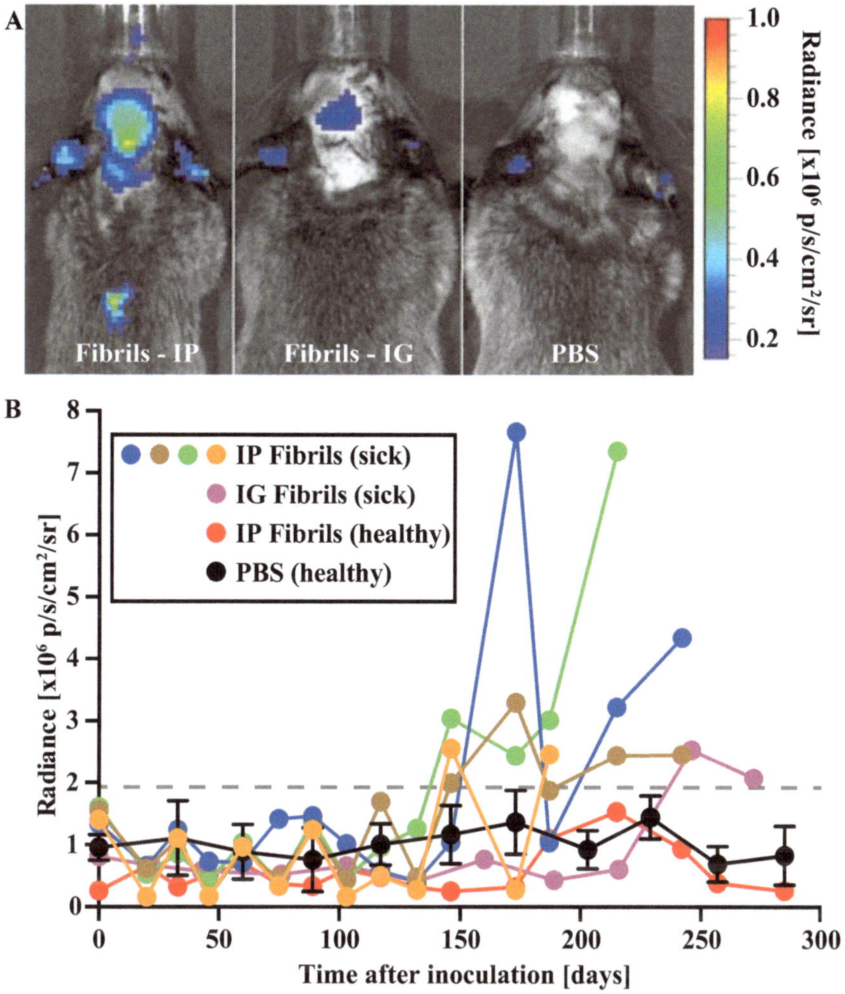

Fig. 3 Gliosis in the CNS of Tg(M83$^{+/-}$:*Gfap*-luc$^{+/-}$) mice after peripheral challenge with alpha-synuclein fibrils was measured by bioluminescence imaging. (**a**) Bioluminescence imaging showed enhanced radiance from the CNS of Tg(M83$^{+/-}$:*Gfap*-luc$^{+/-}$) mice that had been intraperitoneally injected with alpha-synuclein fibrils (left panel), which was caused by an induction of astrocytic gliosis, shortly before the mice developed signs of neurologic disease. The basal radiance in the CNS of PBS-injected control mice did not increase with time (right panel). After intraglossal inoculation with alpha-synuclein fibrils, mouse 121 showed signs of astrocytic gliosis before it died at 285 days (center panel). (**b**) After intraperitoneal challenge with alpha-synuclein fibrils, increased levels of bioluminescence ($>2 \times 10^6$ p/s/cm^2/sr) were measured from the brains of four Tg(M83$^{+/-}$:*Gfap*-luc$^{+/-}$) mice (blue, green, brown, and orange circles) several weeks before they developed neurologic signs of disease. One of five animals also showed elevated levels of bioluminescence shortly before it died 285 days after intraglossal injection with alpha-synuclein fibrils (magenta circles). The measured radiance for healthy, PBS-injected control mice (black circles; error bars show SD [$n = 4$]) and one mouse that did not develop disease after intraperitoneal injection with alpha-synuclein fibrils (red circles) always remained below a threshold of 2×10^6 p/s/cm^2/sr. Bar = 20 μm

2.1 Animal Model

1. Hemizygous Tg(*Gfap*-luc$^{+/-}$) mice [20, 21].
2. Hemizygous Tg(M83$^{+/-}$) or B6;C3-Tg(*Prnp*-SNCA*A53T) 83Vle/J mice [19].

2.2 Inoculum Preparation

1. Tris-buffered saline (TBS): 1× final solution in water.
2. Recombinant alpha-synuclein protein fibrils: prepare monomeric recombinant human alpha-synuclein with the A53T mutation in Tris-buffered saline (TBS) containing 150 mM NaCl in 20 mM Tris–HCl (pH 7.2) [17].
3. Orbital thermomixer.
4. Phosphate-buffered saline (PBS): 1× final solution in water.
5. Sonoplus Mini20 rod sonicator.

2.3 Inoculation

1. Ketamine (100 mg/kg body weight) in isotonic saline solution 0.9%.
2. Xylazine (10 mg/kg body weight) in isotonic saline solution 0.9%.
3. Isoflurane/oxygen anesthetic induction chamber.
4. Standard and narrow pattern forceps.
5. Blunt-nosed thumb forceps with a serrated tip.
6. 27-Gauge disposable hypodermic syringe (*see* **Note 1**).
7. Vet ointment.
8. Heating pad for small rodents.

2.4 Bioluminescence Imaging

1. Isoflurane/oxygen anesthetic induction chamber.
2. Hair shaver.
3. Hair removal cream.
4. Non-skin irritating marker in black.
5. Phosphate-buffered saline (PBS): 1× final solution in water.
6. D-luciferin potassium salt: prepare at a concentration of 30 mg/mL in 1× PBS. Prepare 1 mL aliquots and store them at −20 °C. Protect from light.
7. Weighing scale.
8. IVIS Lumina II imaging system (PerkinElmer).
9. Living Image 3.0 Software (PerkinElmer).

2.5 Immunofluorescence Analysis

Paraffin drops.

1. Fully automated rotary microtome.
2. SuperFrost Ultra Plus microscope slides.
3. Xylol solution 99%.

4. Ethanol: Use a stock of 100%. Prepare different grades of water-diluted ethanol (90%, 70%, and 50%).
5. Citrate buffer (pH 6.0): Combine 9 mL of 0.1 M citric acid and 41 mL of 0.1 M sodium citrate dehydrate. Complete volume with 450 mL water (*see* **Note 2**). Measure the pH before use.
6. Blocking buffer: 20% normal goat serum (*see* **Note 3**), 1% bovine serum albumin (*see* **Note 3**), 1× PBS, 0.5% Triton X-100. Complete volume with water (*see* **Note 4**).
7. Primary antibody buffer: 1% normal goat serum (*see* **Note 3**), 1% bovine serum albumin (*see* **Note 3**), 1× PBS, 0.25% Triton X-100. Complete volume with water (*see* **Note 4**).
8. Secondary antibody buffer: 1% normal goat serum (*see* **Note 3**), 1% bovine serum albumin (*see* **Note 3**), 1× PBS. Complete volume with water (*see* **Note 4**).
9. Primary antibodies: anti-glial fibrillary acidic protein (GFAP) from rabbit. Working dilution: 1:200. Anti-IBA-1 from rabbit. Working dilution: 1:500. Anti-alpha-synuclein, phospho S129 [pSyn#64] from mouse. Working dilution: 1:1200.
10. Secondary antibodies (*see* **Note 5**): Alexa Fluor 488-conjugated anti-rabbit and Alexa Fluor 594-conjugated anti-mouse. Working dilution: 1:1000.
11. 4,6-diamidino-2-phenylindole (DAPI). Working dilution 1:50,000.
12. Fluoromount media.
13. Cover slides.
14. LSM700 confocal laser scanning microscope (Carl Zeiss), or other similar microscope system.

3 Methods

Ensure that all procedures involving animals are performed with approval of your institution's animal care and use committee. House and care for animals according to standard conditions with a 12 h light/dark cycle and free access to food and water. Carry out all procedures at room temperature unless otherwise specified.

3.1 Animal Model

1. Intercross hemizygous Tg(*Gfap*-luc$^{+/-}$) mice with hemizygous Tg(M83$^{+/-}$) mice to generate hemizygous bigenic Tg(M83$^{+/-}$:*Gfap*-luc$^{+/-}$) mice (*see* **Note 6**).
2. Genotype the progeny with real-time PCR for the presence of the transgene encoding human alpha-synuclein and with standard PCR for firefly luciferase [17].
3. Inoculate the animals at an age of 6–8 weeks (*see* **Note 7**).

3.2 Inoculum Preparation

1. Prepare fibrillar alpha-synuclein for inoculations by agitating 3 μg/μL of monomeric recombinant human alpha-synuclein in an orbital shaker at 800 rpm and 37 °C for 5 days [9].
2. Dilute fibril assemblies in phosphate-buffered saline (PBS) to reach a final concentration of 1 μg/μL for intraperitoneal or 2 μg/μL for intraglossal inoculations.
3. Fragment the alpha-synuclein fibrils with a rod sonicator for 1 min (40 pulses of 0.5 s duration with a 1 s pause between each pulse and an amplitude set to 50%) on ice (*see* **Notes 8** and **9**).

3.3 Intraglossal Injections

1. Anesthetize animals with an intraperitoneal injection of 100 mg/kg ketamine and 10 mg/kg xylazine (*see* **Notes 10** and **11**).
2. Fix narcotized animals carefully with adhesive tape onto a heating pad at 37 °C in a dorsal recumbent position with their heads facing toward the investigator.
3. Fill a 27-gauge disposable hypodermic syringe with 5 μL of sonicated alpha-synuclein fibrils or PBS (*see* **Note 12**).
4. Use a pair of blunt-nosed thumb forceps with serrated tips to hold the animal's mouth open, and a second smaller pair of forceps to carefully pull out the tongue to make the bottom side of the tongue accessible for injection.
5. Insert the needle of the syringe into the right or left bottom side of the tongue in proximity to the hypoglossal nerve (*see* **Note 13**). Slowly inject the inoculum after 5 s. Slowly retract the needle after 5 s to ensure that the inoculum has penetrated into the tissue and is not lost while retracting the needle (*see* **Note 14**).
6. Release the animal from its fixations and leave it on the heating plate at 37 °C under constant monitoring until complete recovery.

3.4 Intraperitoneal Injections

1. For intraperitoneal injections, narcotize the animals shortly in an anesthesia chamber with isoflurane/oxygen using a flow rate of 2 L/min and the vaporizer set to 2% (*see* **Notes 10** and **11**).
2. Fill a 27-gauge disposable hypodermic syringe with 50 μL of sonicated alpha-synuclein fibrils or PBS (*see* **Note 12**).
3. Directly inject into the peritoneum of the mouse and avoid penetrating the small intestine or cecum located behind the abdominal wall by holding the animal in a dorsal position with the head facing away from the investigator and downward at approximately 45° (*see* **Note 15**).
4. Leave the animal on the heating pad at 37 °C and monitor the animal until full recovery from anesthesia.

3.5 Bioluminescence Imaging

3.5.1 Picture Acquisition

1. Image Tg(M83$^{+/-}$:*Gfap*-luc$^{+/-}$) mice using a bioluminescence imaging system (*see* **Note 16**).
2. Prior to imaging shortly anesthetize mice in an isoflurane/oxygen chamber with a flow rate of 2 L/min and the vaporizer set to 2% (*see* **Notes 10** and **11**).
3. Shave and depilate the heads of the mice with a depilatory cream (*see* **Note 17**).
4. Weigh the animals prior to injection. Calculate the exact volume of D-luciferin for injection of 150 mg/kg body weight. Intraperitoneally inject D-luciferin (*see* **Notes 18** and **19**) and return the animal to the anesthesia chamber.
5. Start the imaging software. After the imaging system has reached its operating temperature initialize the system by clicking on the 'Initialize' button in the control panel.
6. Select the buttons 'Luminescent' and 'Photograph'. Set the exposure time to 60 s, binning to 'Medium', F/Stop to '1', and the EM gain to 'Off'.
7. Confirm that the excitation filter is set to 'Block' and the emission filter to 'Open', and set the subject height to 1.50 cm.
8. Ten minutes after injection with D-luciferin, place the animals onto the heating plate in the imaging chamber and ensure that their muzzles are correctly placed in the anesthesia outlet. Close the isoflurane/oxygen flow from the inhalation chamber and open the flow for the imaging chamber with a flow rate of 0.25 L/min and an evaporation of 2%. Close the door of the imaging chamber properly.
9. Click on the 'Acquire' button to measure the bioluminescence, which takes 60 s.
10. Stop the isoflurane flow and monitor the animals until they have completely recovered after returning them back to their cages.

3.5.2 Picture Analysis

1. Use the imaging software to quantify the bioluminescence. Within the 'Tool Palette' select 'ROI Tools'. Under 'Type' within the 'ROI Tools' select 'Measurement ROI'.
2. Within the 'ROI Tools' click the 'circle' tool and select the number of ROIs to draw from the pull down menu—ideally '3' for three mice.
3. In the image, right-click each ROI and under 'Properties' adjust the width and height to 1.25 cm. Position each ROI over the brain area that is quantified.
4. In the upper left of the image, set the units panel to 'radiance (photons)'.

5. Within the 'Tool Palette', select 'Image Adjust' and adjust the minimum and the maximum of the 'Color Scale', for instance from 0.20e6 to 1.00e6.
6. Under 'File' save the data with 'Save'.

3.6 Immunofluorescence Analysis

1. Perfuse the animals with saline solution and 4% formalin.
2. Dehydrate the dissected brains and spinal cords in a tissue processing station and embed them in paraffin using a paraffin station.
3. Cut the paraffin-embedded tissues with a microtome in 8-μm thick coronal sections and mount the sections on glass slides.
4. Deparaffinize the tissue sections by incubating them in two separate xylol baths for 5–10 min and rehydrate them through a series of graded ethanol baths (100%, 90%, 70%, and 50%) and finally with water.
5. Incubate the sections in 0.01 M citrate buffer (pH 6.0) for 5 min at room temperature and additionally boil them for 10 min in a microwave oven (*see* **Note 2**).
6. Block the tissue sections by incubation in blocking buffer for 1 h at room temperature.
7. Incubate the sections with the primary antibody overnight at 4 °C.
8. Wash the sections twice with 0.25% (vol/vol) Triton X-100 in 1× PBS and once with 1× PBS.
9. Stain the sections with the corresponding fluorophore-conjugated secondary antibodies and the nuclear dye DAPI diluted in secondary buffer for 1 h at room temperature (*see* **Note 5**).
10. After washing twice with 0.25% (vol/vol) Triton X-100 in 1× PBS and once with 1× PBS, coverslip the slides with embedding media and visualize the staining with a confocal laser scanning microscope.

4 Notes

1. To avoid bubble formation during inoculation, it is better to work with small volume (not more than 300 μL) syringes.
2. Stock solutions of 0.1 M citric acid buffer and 0.1 M sodium citrate dehydrate buffer can be prepared in advance and kept at 4 °C. The working solution of citrate buffer pH 6.0 has to be freshly prepared each time before use.
3. Filter the solution with a 0.22 μm filter and prepare 1 mL aliquots. Store them at −20 °C.

4. Always prepare fresh blocking, primary, and secondary buffer before use.
5. Centrifuge the secondary antibody mix before use at 14,000 × *g* for 1 h at 4 °C. This step precipitates small debris originating from the secondary antibody solution.
6. The use of Tg(M83$^{+/-}$:*Gfap*-luc$^{+/-}$) mice enables noninvasive visualization of astrogliosis and represents a simple method of monitoring inflammatory responses caused by the neuroinvasion of alpha-synuclein over the entire life span of the mouse.
7. The *Gfap*-luc transgene only allows detection of astrogliosis but not of microgliosis, which is equally important in neuroinflammation.
8. Seemingly trivial factors like aggregation time, sonication time, the abundance of endotoxins, or the amount of fibrils used for injection can influence the outcome of an experiment by affecting the seeding propensity of misfolded alpha-synuclein.
9. To avoid contamination by cross-seeding use clean sonication probes to prepare different inocula.
10. Pinch the animal's toe and confirm that it does not withdraw its hind limb to ensure proper anesthesia.
11. Use vet ointment on the animal's eyes to prevent dryness while under anesthesia.
12. A limitation of this protocol is that based on the site of injection the injected inoculum cannot exceed a certain volume. Thus, injections into the tongue can only be performed with smaller volumes than those into the peritoneum, which may affect transmission times to the CNS.
13. For intraglossal injections it has to be considered that not only the hypoglossal nerve but also other cranial nerves like the glossopharyngeal nerve innervate the tongue and could be involved in the interneuronal transport of pathological alpha-synuclein to the brain. Thus, targeting different areas of the tongue could result in different spreading kinetics to the brain. We did not directly inject into the hypoglossal nerve and it is possible that targeting the hypoglossal nerve could improve CNS transmission of pathological alpha-synuclein.
14. A fast retraction of the needle after intraglossal inoculation can cause the loss of sample or bleeding of the tongue.
15. For intraperitoneal injections with alpha-synuclein fibrils it is critical that the inoculum is correctly injected into the peritoneum and not accidentally into internal organs like the intestine or cecum, which could negatively affect neuroinvasion.
16. We imaged mice every 2–4 weeks but alternative imaging intervals are possible.

17. Color the ears in black with a nonirritating marker to block unspecific bioluminescence.
18. To obtain uniform results during imaging it is also critical to always freshly prepare the D-luciferin solution prior to injection and to let it exactly incubate for 10 min after intraperitoneal injection before imaging.
19. Mistargeting of D-luciferin to internal organs could slow down its distribution to the brain and result in lower bioluminescence.

References

1. Tamgüney G, Korczyn AD (2017) A critical review of the prion hypothesis of human synucleinopathies. Cell Tissue Res 373(1):213–220. https://doi.org/10.1007/s00441-017-2712-y
2. Kordower JH, Chu Y, Hauser RA, Freeman TB, Olanow CW (2008) Lewy body-like pathology in long-term embryonic nigral transplants in Parkinson's disease. Nat Med 14(5):504–506. https://doi.org/10.1038/nm1747
3. Li JY, Englund E, Holton JL, Soulet D, Hagell P, Lees AJ, Lashley T, Quinn NP, Rehncrona S, Bjorklund A, Widner H, Revesz T, Lindvall O, Brundin P (2008) Lewy bodies in grafted neurons in subjects with Parkinson's disease suggest host-to-graft disease propagation. Nat Med 14(5):501–503. https://doi.org/10.1038/nm1746
4. Li W, Englund E, Widner H, Mattsson B, van Westen D, Latt J, Rehncrona S, Brundin P, Bjorklund A, Lindvall O, Li JY (2016) Extensive graft-derived dopaminergic innervation is maintained 24 years after transplantation in the degenerating parkinsonian brain. Proc Natl Acad Sci U S A 113(23):6544–6549. https://doi.org/10.1073/pnas.1605245113
5. Hansen C, Angot E, Bergstrom AL, Steiner JA, Pieri L, Paul G, Outeiro TF, Melki R, Kallunki P, Fog K, Li JY, Brundin P (2011) Alpha-Synuclein propagates from mouse brain to grafted dopaminergic neurons and seeds aggregation in cultured human cells. J Clin Invest 121(2):715–725. https://doi.org/10.1172/JCI43366
6. Mougenot AL, Bencsik A, Nicot S, Vulin J, Morignat E, Verchere J, Betemps D, Lakhdar L, Legastelois S, Baron TG (2011) Transmission of prion strains in a transgenic mouse model overexpressing human A53T mutated alpha-synuclein. J Neuropathol Exp Neurol 70(5):377–385. https://doi.org/10.1097/NEN.0b013e318217d95f
7. Luk KC, Kehm VM, Zhang B, O'Brien P, Trojanowski JQ, Lee VM (2012) Intracerebral inoculation of pathological alpha-synuclein initiates a rapidly progressive neurodegenerative alpha-synucleinopathy in mice. J Exp Med 209(5):975–986. https://doi.org/10.1084/jem.20112457
8. Luk KC, Kehm V, Carroll J, Zhang B, O'Brien P, Trojanowski JQ, Lee VM (2012) Pathological alpha-synuclein transmission initiates Parkinson-like neurodegeneration in nontransgenic mice. Science 338(6109):949–953. https://doi.org/10.1126/science.1227157
9. Masuda-Suzukake M, Nonaka T, Hosokawa M, Oikawa T, Arai T, Akiyama H, Mann DM, Hasegawa M (2013) Prion-like spreading of pathological alpha-synuclein in brain. Brain 136(Pt 4):1128–1138. https://doi.org/10.1093/brain/awt037
10. Watts JC, Giles K, Oehler A, Middleton L, Dexter DT, Gentleman SM, DeArmond SJ, Prusiner SB (2013) Transmission of multiple system atrophy prions to transgenic mice. Proc Natl Acad Sci U S A 110(48):19555–19560. https://doi.org/10.1073/pnas.1318268110
11. Recasens A, Dehay B, Bove J, Carballo-Carbajal I, Dovero S, Perez-Villalba A, Fernagut PO, Blesa J, Parent A, Perier C, Farinas I, Obeso JA, Bezard E, Vila M (2014) Lewy body extracts from Parkinson disease brains trigger alpha-synuclein pathology and neurodegeneration in mice and monkeys. Ann Neurol 75(3):351–362. https://doi.org/10.1002/ana.24066
12. Holmqvist S, Chutna O, Bousset L, Aldrin-Kirk P, Li W, Bjorklund T, Wang ZY, Roybon L, Melki R, Li JY (2014) Direct evidence of Parkinson pathology spread from the gastrointestinal tract to the brain in rats. Acta Neuropathol 128(6):805–820. https://doi.org/10.1007/s00401-014-1343-6

13. Sacino AN, Brooks M, Thomas MA, McKinney AB, Lee S, Regenhardt RW, McGarvey NH, Ayers JI, Notterpek L, Borchelt DR, Golde TE, Giasson BI (2014) Intramuscular injection of alpha-synuclein induces CNS alpha-synuclein pathology and a rapid-onset motor phenotype in transgenic mice. Proc Natl Acad Sci U S A 111(29):10732–10737. https://doi.org/10.1073/pnas.1321785111
14. Bernis ME, Babila JT, Breid S, Wüsten KA, Wüllner U, Tamgüney G (2015) Prion-like propagation of human brain-derived alpha-synuclein in transgenic mice expressing human wild-type alpha-synuclein. Acta Neuropathol Commun 3(1):75. https://doi.org/10.1186/s40478-015-0254-7
15. Peelaerts W, Bousset L, Van der Perren A, Moskalyuk A, Pulizzi R, Giugliano M, Van den Haute C, Melki R, Baekelandt V (2015) Alpha-Synuclein strains cause distinct synucleinopathies after local and systemic administration. Nature 522(7556):340–344. https://doi.org/10.1038/nature14547
16. Prusiner SB, Woerman AL, Mordes DA, Watts JC, Rampersaud R, Berry DB, Patel S, Oehler A, Lowe JK, Kravitz SN, Geschwind DH, Glidden DV, Halliday GM, Middleton LT, Gentleman SM, Grinberg LT, Giles K (2015) Evidence for alpha-synuclein prions causing multiple system atrophy in humans with parkinsonism. Proc Natl Acad Sci U S A 112(38):E5308–E5317. https://doi.org/10.1073/pnas.1514475112
17. Breid S, Bernis ME, Babila JT, Garza MC, Wille H, Tamgüney G (2016) Neuroinvasion of alpha-synuclein prionoids after intraperitoneal and intraglossal inoculation. J Virol 90(20):9182–9193. https://doi.org/10.1128/JVI.01399-16
18. Bartz JC, Kincaid AE, Bessen RA (2003) Rapid prion neuroinvasion following tongue infection. J Virol 77(1):583–591. https://doi.org/10.1128/jvi.77.1.583-591.2003
19. Giasson BI, Duda JE, Quinn SM, Zhang B, Trojanowski JQ, Lee VM (2002) Neuronal alpha-synucleinopathy with severe movement disorder in mice expressing A53T human alpha-synuclein. Neuron 34(4):521–533. https://doi.org/10.1016/S0896-6273(02)00682-7
20. Zhu L, Ramboz S, Hewitt D, Boring L, Grass DS, Purchio AF (2004) Non-invasive imaging of GFAP expression after neuronal damage in mice. Neurosci Lett 367(2):210–212. https://doi.org/10.1016/j.neulet.2004.06.020
21. Tamgüney G, Francis KP, Giles K, Lemus A, DeArmond SJ, Prusiner SB (2009) Measuring prions by bioluminescence imaging. Proc Natl Acad Sci U S A 106(35):15002–15006. https://doi.org/10.1073/pnas.0907339106

Chapter 11

Continual Conscious Bioluminescent Imaging in Freely Moving Mice

Juan Antinao Diaz, Amy Geard, Lorna M. FitzPatrick, Juliette M. K. M. Delhove, Suzanne M. K. Buckley, Simon N. Waddington, Tristan R. McKay, and Rajvinder Karda

Abstract

In vivo bioluminescent imaging allows the detection of reporter gene expression in rodents in real time. Here we describe a novel technology whereby we can generate somatotransgenic rodents with the use of a viral vector carrying a luciferase transgene. We are able to achieve long term luciferase expression by a single injection of lentiviral or adeno-associated virus vectors to newborn mice. Further, we describe whole body bioluminescence imaging of conscious mice in a noninvasive manner, thus enforcing the 3R's (replacement, reduction, and refinement) of biomedical animal research.

Key words Lentiviral vectors, Adeno-associated vectors, Transcription factor activated reporter, Biosensors, Luciferase, Neonatal mice, Luciferin, Bioluminescence imaging, Conscious imaging

1 Introduction

Bioluminescence imaging has been established for many years with the use of germline transgenics. In germline light producing transgenic (LPT) models the luciferase transgene is inserted downstream of an endogenous gene promoter and thus results in every cell carrying a copy of the luciferase reporter gene [1, 2]. The advantage of this technology is that it provides a whole-body readout of the specific promoter or enhancer activity. However, monitoring bioluminescence from individual organs and tissues has proven to be difficult due to the "background noise" generated by the whole-body bioluminescence. Furthermore, LPTs are produced by frequent backcrossing, which results in increased time, cost and number of animals.

We have developed a novel technology whereby we can deliver a luciferase reporter gene using viral vectors (lentivirus and adeno-associated virus (AAV)), directly to the targeted tissue of newborn

Steven Ripp (ed.), *Bioluminescent Imaging: Methods and Protocols*, Methods in Molecular Biology, vol. 2081,
https://doi.org/10.1007/978-1-4939-9940-8_11,

rodents, to generate somatotransgenic animals. We have previously shown that viral vector delivery to neonatal mice confers immune tolerization to the transgenic protein [3] and long-term gene expression [4]. Viral vectors can be engineered for cell or tissue specificity by pseudotyping or by capsid engineering, with specific viral envelopes or capsids. The development of highly sensitive charge coupled device (CCD) cameras has allowed the continual detection of luciferase activity in a noninvasive manner on live mice [5].

Previous studies have shown that prolonged exposure to isoflurane can cause primary neuronal axon outgrowth [6], microglia activation, cell death and disruption of neurodevelopment [7]. We therefore interrogated the use of somatotransgenic bioimaging technology as a means of quantifying bioluminescent imaging on conscious moving mice [8]. Using a mouse model of cerebral palsy, we monitored inflammation in a noninvasive manner, and established that there was no difference in luciferase expression between conscious and unconscious mice. This demonstrates the utility of the methodology and is conducive to implementation of the 3R's of biomedical animal research to replace or reduce animal numbers or refine methods involving their use [8].

2 Materials

2.1 Plasmids

Both lentiviral and AAV biosensor plasmids were designed to be compatible with Gateway® cloning technology. The plasmids contain a Gateway® recombination cassette, upstream of a 3× FLAG-tagged tag, codon-optimized luciferase reporter gene, linked by a T2A bicistronic linker and an enhanced green fluorescent protein (GFP), referred to as pLNT-GW-JDG and AAV-GW-JDG (Fig. 1) [1, 9]. Entry plasmids were also designed, which contained the promoter or transcription factor response element of choice (Fig. 1) [10]. Plasmids should always be dissolved in sterile TE buffer or in nuclease free water.

1. Lentiviral plasmid pLNT-GW-JDG.
2. AAV plasmid AAV-GW-JDG (synthesized from Aldevron).
3. Entry plasmid pENTR (Aldevron) containing response element of choice.

2.2 Cloning

1. LR Clonase kit (Invitrogen).
2. DH5α chemically competent cells.
3. Stbl3 or NEB Stable chemically competent cells.
4. *Bam*HI and *Eco*RI restriction enzymes.

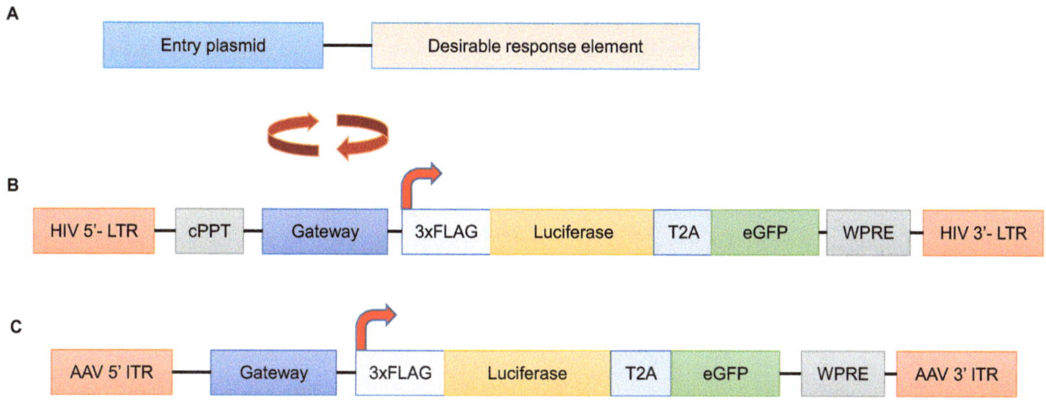

Fig. 1 Schematic diagram of the constructs used in Gateway cloning. (**a**) The pENTR plasmid contains the desired response element or promoter. (**b**) Lentiviral construct contains the central polypurine tract (cPPT), gateway upstream of a 3× FLAG-tagged tag, codon optimized luciferase, linked to enhanced green fluorescent protein by a T2A sequence. Also contains a woodchuck hepatitis virus posttranscriptional regulatory element (WPRE). (**c**) From 5′ITR to 3′ITR the AAV construct contains the same sequence as the lentiviral plasmid, except for the cPPT element. Not drawn to scale

5. Appropriate restriction enzyme that can be used to screen clones.
6. LB broth: 5 g/L yeast extract, 5 g/L NaCl, 10 g/L peptone, 5 mM NaOH.
7. LB agar: LB medium containing 14 g/L agar.

2.3 Lentiviral Production: Plasmids

Plasmids should always be dissolved in sterile TE buffer or in nuclease free water.

1. Envelope plasmid VSV-G (pMD2.G).
2. Packaging plasmid gag-pol, tat, rev plasmid (pCMVΔR8.74).
3. Payload plasmid pLNT-Response element-JDG.

2.4 Lentiviral Production: Cell Culture and Transfection

1. HEK293T cells (Biobank at Institute of Child Health, University College London).
2. 37 °C cell culture incubator with 5% CO_2 atmosphere.
3. DMEM supplemented with 10% fetal bovine serum and 1% Pen/Strep.
4. Tissue culture flasks, T175.
5. Opti-MEM.
6. Trypsin-EDTA.
7. Phosphate buffered saline (PBS).
8. 0.45 μm PVDF membrane filters.
9. Polyethylenimine (PEI) Max MW 40,000 (Polysciences Inc).

2.5 AAV Production: Plasmids

Plasmids should always be dissolved in sterile TE buffer or in nuclease free water.

1. Helper plasmid pHGTI.
2. Packaging plasmid, in this chapter pAAV2/8.
3. Payload plasmid AAV-Response element-JDG.

2.6 AAV Production: Cell Culture and Transfections

1. AAV pro 293T cells (Takara Bio).
2. 37 °C cell culture incubator with 5% CO_2 atmosphere.
3. DMEM supplemented with 10% fetal bovine serum and 1% Pen/Strep.
4. 15 cm Tissue culture dishes.
5. Opti-MEM.
6. Cell scrapers.
7. 0.22 μm and 0.45 μm PVDF membrane filters.
8. PEI MW 40,000.
9. 0.1% PEI solution: 1 g of PEI MW 40,000, 900 mL of dH_2O, pH to 7.2, make up to 1 L
10. TD Buffer: 140 mM NaCl (MW 58.44; 8.181 g/L), 5 mM KCl (MW 74.56; 372.8 mg/L), 0.7 mM K_2HPO_4 (MW 174,18; 121.9 mg/L), 3.5 mM $MgCl_2$ (MW 95.21; 333.23 mg/L), 25 mM Tris (MW 121.14; 3.0285 g/L), adjust pH to 7.5, autoclave, fill to 1 L with dH_2O.
11. Glycine buffer (100 nM): 3.75 g of glycine (100 mM final concentration), 900 mL of dH_2O, adjust to pH 2, fill to 1 L with dH_2O, filter at 0.22 μM.
12. Tris buffer (1 M): 12.114 g of Tris base (final concentration 1 M), 80 mL of dH_2O, adjust solution to pH 8.8, add dH_2O to final volume of 100 mL.
13. 10% sodium deoxycholate solution: 1 g of sodium deoxycholate, 10 mL of dH_2O.
14. Ethanol.
15. PBS.
16. Benzonase.
17. POROS™ CaptureSelect™ AAVX Resin or prepacked column.
18. Dialysis cassette (Side-A-Lyzer; ThermoFisher Scientific).
19. Amicon Ultra 15_100kDa MWCO.
20. 1.5 mL Eppendorf with 0.22 μm filter.

Table 1
Primers and probes used for lentivirus and AAV vector titration

Primer	Sequence
LTR—forward	5′-TGTGTGCCCGTCTGTTGTGT-3′
LTR—reverse	5′-GAGTCCTGCGTCGAGAGAGC-3′
LTR—probe	5′-(FAM)-CAGTGGCGCCCGAACAGGGA-(3BHQ_1)-3′
β-actin—forward	5′-GTAGCACAGCTTCTCCTTAAT-3′
β-actin—reverse	5′-GGACCTGACTGACTACCT-3′
β-actin—probe	5′-(FAM)-CGCGCTCGGTGAGGATCTTCAT-(3BHQ_1)-3′
SV40—forward	5′-CACAACTAGAATGCAGTGAAA-3′
SV40—reverse	5′-AACTTGTTTATTGCAGCTTAT-3′
SV40—probe	5′-(FAM)-TGGTTACAAATAAAGCAATAGCATCAC-(3BHQ_1)-3′

2.7 Titration of Lentiviral and AAV Vectors

1. Blood and Tissue Kit (Qiagen).
2. StepOne Plus, QuantStudio 3 (ThermoScientific) or similar qPCR machine.
3. Primers and probe set for target sequences (Table 1), LTR [11] and β-actin are used for lentivirus and SV40 for AAV titration.
4. Standard for qPCR; plasmid used to produce the vector (β-actin should be present in the construct). Prepare a serial dilution (1:10) between 1×10^9 copies/5 μL down to 1×10^2 copies/5 μL using dilution buffer. The range can be adjusted according to your particular vector.
5. Luna Universal Probe qPCR MasterMix (NEB).
6. Nuclease-free water (NEB).
7. Pluronic F-68 nonionic surfactant (100×).
8. Dilution buffer: dilute Pluronic F-68 (100×) to 1:100 in water.
9. 96-well PCR plate 0.1 mL format.
10. MicroAmp® Optical Adhesive Film (Applied Biosystems).

2.8 In Vivo Bioluminescent Imaging

1. Potassium luciferin, D-luciferin.
2. Dulbecco's phosphate-buffered saline (DPBS), no calcium and magnesium.
3. 33-gauge Hamilton needle.
4. 1 mL syringe with needle (16 mm).
5. Imaging chamber box, 5 cm × 5 cm × 6 cm.
6. Perspex box, as shown in Karda et al. [8].

7. IVIS imaging system (Perkin Elmer).
8. Oxygen.
9. Isoflurane.
10. Mice CD1 strain (Charles River Laboratories, UK).

3 Methods

3.1 Cloning Lentiviral and AAV Plasmids

Cloning of synthetic promoter and response elements in lentiviral plasmids has been previously described in greater detail [10].

1. Design serial transcription factor binding sequence (TFBS) with 4–10 interspaced binding sequences with 10 random nucleotides between each TFBS.
2. Synthesize TFBS sequence with flanking 5′ *Bam*HI and 3′ *Eco*RI restriction sites. Use for direct cloning into pENTR.
3. *Bam*HI/*Eco*RI digest the pENTR vector and gel extract backbone.
4. Clone TFBS sequence into pENTR, transform into DH5α competent cells, and select using kanamycin-laden agar plates (50 μg/mL).
5. Select individual clones, grow in kanamycin-laden LB broth (50 μg/mL) and isolate plasmid.
6. Perform a *Bam*HI/*Eco*RI double digest to confirm presence of the TFBS insert in the pENTR plasmid.
7. Confirm sequence using the following pENTR primer. The primer binding site is 81 bp upstream from the *Bam*HI site (ACTGATAGTGACCTGTTCGTTGC).
8. Using the LR clonase kit, perform the LR reaction between the pENTR plasmid containing the response element or promoter of choice and pLNT-GW-JDG or AAV-GW-JDG, by following the manufacturer's guidelines.
9. Transform 1 μL of reaction mixture into Stbl3 competent cells and grow overnight at 30 °C on an ampicillin (100 μg/mL) LB-agar plate.
10. Select individual colonies and grow in shaking incubator (200 rpm) overnight at 30 °C in 5 mL of LB broth containing ampicillin (100 μg/mL).
11. Screen colonies using an appropriate restriction enzyme digest to distinguish the clones form the original Gateway-containing parental plasmid and perform gel electrophoresis.
12. For pLNT-GW-JDG, it is a *BamHI* digest (10, 928 bp, 857 bp, 702 bp, 228 bp, 12 bp (not seen on gel)). The 857 and 702 bands are removed and exchanged for an

approximately 300 bp band when recombination has been successful. For AAV-GW-JDG it is dependent on the insert.

3.2 Lentivirus Transfection

Carry out all experiments in a cell culture hood.

1. Seed 2×10^7 HEK293T cells in T175 flasks and leave overnight until they are 90% confluent (*see* **Note 1**).
2. For each flask mix the following concentrations of plasmids in 6 mL of Opti-MEM; 50 μg of payload plasmid, containing luciferase transgene, 32.5 μg of packaging plasmid and 17.5 μg of envelop plasmid. In a separate falcon tube, mix 1 μL of 10 mM PEI in 6 mL of Opti-MEM.
3. Mix the two plasmid DNA and PEI solutions together and leave to incubate for 20 min at room temperature.
4. Remove DMEM from cells and wash with DPBS.
5. Add 12 mL of the PEI and plasmid DNA mix to each flask and incubate for 4 h. After which remove media and replace with fresh 15 mL of DMEM (supplemented with 10% FBS and 1% Pen/Strep).
6. After 48 h remove supernatant and filter in 0.45 μm PVDF membrane. Add supernatant to 50 mL falcon tubes and spin for 20 h at $4800 \times g$ at 4 °C.
7. Remove supernatant (*see* **Note 2**) and add 50 μL of Opti-MEM per flask to tube, in order to resuspend viral pellet. Leave on ice for 1 h with gentle mixing every 20 min, careful not to introduce bubbles and then cryopreserve at −80 °C.

3.3 Lentiviral Titration

The HT1080 cell line was used for this protocol, as they have been shown to have a stable karyotype [12].

1. Seed 100,000 cells into each well of a tissue cultured treated 6-well plate.
2. The following day add the virus; using concentrated vector in dilutions of: 1:2, 1:4, 1:8, 1:16 and 1:32.
3. 72 h post transfection, passage cells. Repeat this once more.
4. Collect cells and extract DNA using Blood and Tissue kit (Qiagen), following manufacturer's protocol.
5. Determine concentration of DNA sample. Each sample should be diluted to 20 ng/μL.
6. All samples are run in triplicates, along with standards.
7. Add the 15 μL master mix, shown in Table 2.
8. Add 5 μL for each sample and standard.
9. Run qPCR using the settings, shown in Table 3.

Table 2
Reaction mixture for lentivirus qPCR

Reagent	Stock concentration	Final concentration	1rx (μL)
Luna qPCR MasterMix	2×	1×	10
Forward primer	100 μM	450 nM	0.09
Reverse primer	100 μM	450 nM	0.09
Probe	100 μM	250 nM	0.05
H_2O	–	–	4.77
Total			15 μL

Table 3
qPCR settings for lentivirus titration

Stage	Temperature	Time	Cycles
Initial denaturation	95 °C	10 min	1
Denaturation	95 °C	15 s	40–45
Extension	60 °C	45 s	

10. Obtain absolute copy number values from qPCR run, based on standard curves of known DNA copy number concentration for LTR and β-actin.
11. Vector titration calculated using formula [13] (*see* **Note 3**):

$$\text{Vector copy number (copies per cell)} = \frac{\text{Copy number LTR}}{\text{Copy number } \beta \text{ actin}} \times 2$$

12. To obtain the functional titer, use the following equation (*see* **Note 4**):

$$\text{Titre (vg/mL)} = \frac{\text{Cells in day 1} \times \text{Vector copy number}}{\text{Volume (mL)}}$$

3.4 AAV Transfection

Carry out all experiments in a cell culture hood.

1. Seed 2×10^7 HEK293T cells in 15 cm dish (using DMEM, 10% FBS and 1% Pen/Strep) and leave overnight until they are 90% confluent (*see* **Note 1**).
2. For each dish mix the following concentrations of plasmids in 1.6 mL of Opti-MEM; 7.14 μg of payload plasmid, containing luciferase transgene, 7.14 μg of AAV-Capsid plasmid and 21.4 μg of Helper plasmid. In a separate falcon tube, mix 75 μL of 0.1% PEI in 1.6 mL of Opti-MEM.

3. Filter the DNA mixture though a 0.22 μm membrane filter.
4. Mix the two plasmid DNA and PEI solutions together.
5. Leave to incubate for 20 min at room temperature to form DNA/PEI complexes.
6. Carefully add 3 mL of the mixture to each dish in a drop-wise fashion.
7. Incubate overnight at 37 °C, 5% CO_2.
8. Remove the media from the cells, replace with 15 mL DMEM, 2% FBS and 1% Pen/Strep.
9. Incubate for 48 h at 37 °C, 5% CO_2.
10. Scrape off cells from each plate and collect with the media into 50 mL falcon tubes and centrifuge at 3100 × *g* for 10 min. Store the supernatant at 4 °C.
11. Resuspend the cell pellet in TD Buffer, 1 mL per plate, and store at −80 °C.
12. Cell pellet: 5 freeze-thaw cycles moving between a water bath at 37 °C and dry ice. Then add 50 μL of 10% sodium deoxycholate per mL of TD buffer and 0.2 μL of Benzonase per mL TD buffer. Incubate for 30 min at 37 °C followed by centrifugation at 3100 × *g* for 30 min at 4 °C. Using a syringe, filter the supernatant through a PES 0.45 μm sterile filter, store the solution at 4 °C for up to 24 h. Discard the pellet.
13. Supernatant: add 1 μL of Benzonase per 10 mL of supernatant and 2 μL of 100 mM magnesium sulfate ($MgSO_4$) per 1 mL of supernatant, incubate for 30 min at 37 °C followed by centrifugation at 3100 × *g* for 30 min at 4 °C. Filter the media through a 0.45 μm PVDF stericup, store the solution at 4 °C for up to 24 h. Discard the pellet.

3.5 AAV Purification by HPLC

1. Clean the HPLC machine with 75 mL of 20% ethanol followed by 75 mL of PBS, make sure the pH and absorbance lines are stable and the values reach baseline (zero is the baseline reading).
2. Position the POROS column and wash with PBS.
3. Position FACS tubes in the system, add 30 μL of 1 M Tris solution to each.
4. Run the supernatant from the cell pellet through the system. The absorbance curve should increase significantly while the solution passes though the column, the pH curve should remain stable. Wash the system with PBS (approximately 75 mL) until the reading reaches the set baseline; collect the waste in another sterile container and run it though the system following the same procedure.
5. Collect the bound vector by eluting into FACS tubes, 1 mL glycine solution. Neutralize the pH by adding 30 μL of 1 M

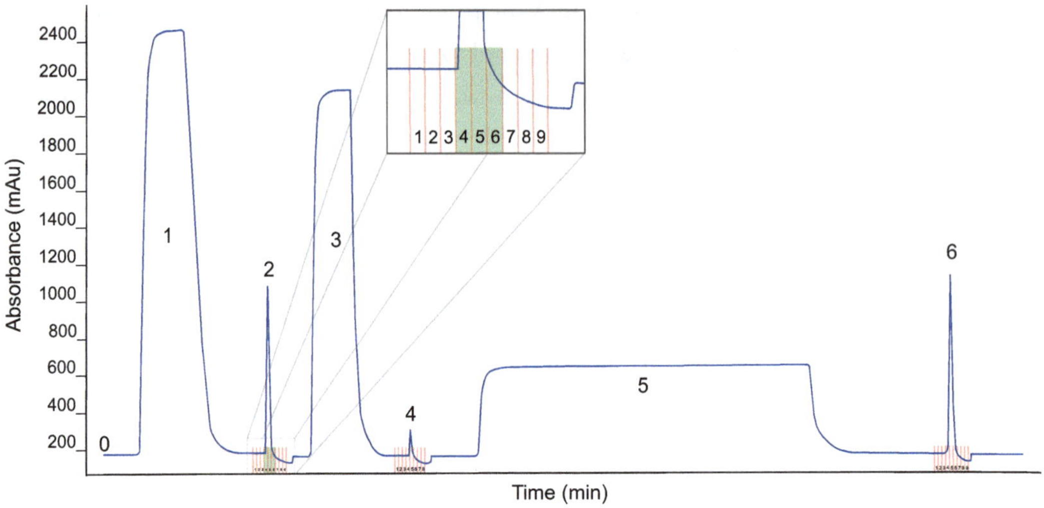

Fig. 2 AAV vector purification by HPLC. The absorbance (in milli-Absorbance Units, mAu) curve (blue) in the HPLC software will show the proteins passing through the system. After establishing a baseline (0), a significant increase in the signal marks the proteins from the cell pellet passing though the column (1). Once baseline is reached again, the vector is collected (2). The insert shows a magnification of peak 2, in which the 1 mL fractions are collected into FACS tubes which contain AAV vector. The fractions which contain vector, show a peak in absorbance. In the figure it is fractions 4, 5 and 6. If waste is collected, it should show a similar increase in the absorbance (3), but the elution peak will be significantly less (4). The supernatant takes longer as the volume is much larger (5). The elution peak reaches similar levels as the one obtained in the cell pellet (6)

Tris to each FACS tube. To identify the fraction containing the vector follow the absorbance curves, the pH should decrease, and absorbance should increase as the fraction is eluted (Fig. 2). The vector could be eluted in more than one fraction.

6. Transfer all the fractions into the dialysis cassette and leave to spin using a magnetic stirrer in 1 L of PBS overnight to restore pH.
7. Wash the system with PBS and repeat the same procedure for the supernatant.

3.6 AAV Concentration

1. Prime the Amicon Ultra 15_100kDa MWCO filter with 5 mL of PBS, centrifuge at 4000 × *g* for 5 min.
2. Add the unconcentrated vector from the cassette to the filter and centrifuge at 4000 × *g* for 5 min.
3. Discard the flow-through and collect the remaining vector in the filter.
4. Add vector to 1.5 mL Eppendorf with 0.22 μm filter, centrifuge at 12,000 × *g* for 3 min.
5. Remove the filter and aliquot and store at 4 °C for short-term use or −80 °C for long-term storage.

3.7 AAV Titration

Digest virus to release DNA:

1. In one PCR tube add: 5 μL of the rAAV vector to be titrated, 5 μL of DNase digestion buffer, 10 units of DNase I, make final volume to 50 μL with dH_2O.
2. Incubate at 37 °C for 30 min and inactivate for 10 min at 65 °C (*see* **Note 5**).
3. Prepare seven serial dilutions of the digested vector, usually 1:10, using the dilution buffer.
4. Prepare qPCR Master Mix (for SV40) (Table 4).
5. Add the Master Mix to all the necessary wells, followed by triplicates of the standards and each dilution of the vector (5 μL of each), for a total volume of 20 μL.
6. For SV40 primers, run the samples with the settings in Table 5.
7. Determine the number of copies obtained for each corresponding dilution from the qPCR machine. Ensure the quality control parameters for the run have been met (*see* **Note 6**).
8. Use the following formula to calculate the vector titration (*see* **Note 7**):

$$\text{Titre (vc/mL)} = \frac{C}{5} \times 1000 \times D \times 10 \times 2$$

Table 4
Master mix for AAV qPCR titration

Reagent	Stock concentration	Final concentration	1rx (μL)
Luna qPCR master mix	2×	1×	10
Forward primer	100 μM	450 nM	0.09
Reverse primer	100 μM	450 nM	0.09
Probe	100 μM	250 nM	0.05
H_2O	–	–	4.77
Total			15 μL

Table 5
qPCR settings for AAV titration

Stage	Temperature	Time	Cycles
Initial denaturation	95 °C	10 min	1
Denaturation	95 °C	15 s	40–45
Extension	60 °C	45 s	

3.8 Animal Procedures

All procedures were performed under United Kingdom Home Office Project License 70/8030, approved by the ethical review committee and followed institutional guidelines as outlined by University College London. Outbred CD1 mice (*see* **Note 8**) were used and supplied by Charles River.

3.8.1 Administration of Vectors to Neonatal Mice

1. New born mice (first day of development), briefly expose to hypothermic anesthesia, by placing them on wet ice.
2. Inject 5 μL of concentrated lentivirus or AAV (*see* **Note 9**) into cerebral lateral ventricles of one brain hemisphere (*see* **Note 10**) using 33 gauge Hamilton needle, by following the coordinates shown by Kim et al. [14].
3. Place mice back in dam.
4. For intravenous injections, briefly expose hypothermic anesthesia. Inject 20 μL into superficial temporal vein [1], using 33 gauge Hamilton needle.
5. Place mice back in dam.

3.8.2 Bioluminescent Imaging

1. Where appropriate (*see* **Note 11**), anesthetize mice with isoflurane and 100% oxygen.
2. Administer 15 mg/mL of D-luciferin to mice via intraperitoneal injections (*see* **Note 12**).
3. Leave for 5 min (*see* **Note 13**).
4. Mice which have received an intracranial injection of vector, should be placed in an imaging chamber with the following dimension; 5 cm × 5 cm × 6 cm (*see* **Note 14**). Mice which have received an intravenous injection, should be placed in Perspex box.
5. Place imaging chambers into IVIS machine and take images between 1 s and 5 min.
6. Return mice to their cages (*see* **Note 15**).
7. Measure luciferase expression by using Living Image Software (Perkin Elmer). Select region of interest (*see* **Note 16**) and measure photons per second per centimeter squared per steradian (photons/second/cm^2/sr).

4 Notes

1. Make sure cells are healthy (mycoplasma negative) and have been passaged every second day for at least three passages prior to vector production.
2. Remove excess media by placing the falcon tube upside down on tissue paper sprayed with 70% IMS.

3. This calculates the vector copy number per cell in the sample; the twofold factor accounts for the two alleles of the β-actin gene.
4. Cell concentrations at the day of transfection are around 200,000. The average titer of all the dilutions from the same sample is considered in the final titer.
5. Optional, add 0.4 units of proteinase K [15], incubate for 1 h at 50 °C and inactivate for 20 min at 95 °C.
6. The qPCR reaction quality should be the following; efficiency between 90 and 110%, $R^2 > 0.98$, replicates within 0.5 Ct of each other, no template control above Ct35.
7. Where C is the copies measured from the reaction. It is divided by 5 to account for the 5 μL used in the qPCR reaction. 1000 is the conversion factor from μL to mL, D is the dilution factor (i.e., for a 1:2 dilution from the digested vector, this will be 2) and 10 is the initial dilution of the AAV vector in the digestion (5 μL in 50 μL). Two account for the complementary strand not targeted by the primers.
8. Mice with white fur are best to use for bioluminescence. Mice with dark fur make detection of the photons difficult as the dark hair absorbs the luminescent signal more so than the white hair. Dark colored mice are therefore routinely shaved in order to maximize luciferase expression.
9. Use concentrated lentivirus that has a qPCR titer greater than 1×10^8 vg/mL. For AAV, a titer greater than 1×10^{11} vg/mL.
10. Keep the needle moist by placing it between a wet tissue paper. This reduces the friction between the needle and skin of mice and prevents the contents of the needle from drying out and blocking the tip.
11. Mice can be imaged consciously, without the use of anesthesia.
12. Use a 33 gauge Hamilton needle to inject mice with D-luciferin, up to 10–12 days of age. After which use a 1 mL syringe and 27 gauge needle. Also make sure that the bladder or other internal organs are not penetrated by the needle.
13. Leave for 5 min after luciferin administration as this allows the luciferin to enter into the bloodstream.
14. Once mice have passed the age of 15/16 days, they should be imaged while conscious in an imaging chamber of the following dimensions; 11 cm × 7 cm × 6 cm.
15. If the mice have been imaged using anesthesia, return mice to cage and wait until they have gained consciousness and are clearly mobile before leaving unattended.
16. The surface area of the imaging chamber is taken as a region of interest. For example, the imaging chamber for intracranial injected mice is 5 cm × 5 cm.

Acknowledgments

J.M.K.M.D. and S.N.W. were funded by the ERC grant Somabio (260862), T.R.M. and S.N.W. were funded by the NC3Rs (NC/L001780/1). R.K. and S.N.W. received funding from MRC grants MR/P026494/1 and MR/R015325/1, and from SPARKS grant 17UCL01. J.A.D. is funded by CONICYT Becas Chile Doctoral Fellowship Program 72160294.

References

1. Buckley SMK, Delhove J, Perocheau DP, Karda R, Rahim AA, Howe SJ, Ward NJ, Birrell MA, Belvisi MG, Arbuthnot P, Johnson MR, Waddington SN, McKay TR (2015) In vivo bioimaging with tissue-specific transcription factor activated luciferase reporters. Sci Rep 5:11842. https://doi.org/10.1038/srep11842
2. Ciana P, Di Luccio G, Belcredito S, Pollio G, Vegeto E, Tatangelo L, Tiveron C, Maggi A (2001) Engineering of a mouse for the *in vivo* profiling of estrogen receptor activity. Mol Endocrinol 15(7):1104–1113. https://doi.org/10.1210/me.15.7.1104
3. Ward NJ, Buckley SMK, Waddington SN, VandenDriessche T, Chuah MKL, Nathwani AC, McIntosh J, Tuddenham EGD, Kinnon C, Thrasher AJ, McVey JH (2011) Codon optimization of human factor VIII cDNAs leads to high-level expression. Blood 117(3):798–807. https://doi.org/10.1182/blood-2010-05-282707
4. Buckley SMK, Howe SJ, Rahim AA, Buning H, McIntosh J, Wong SP, Baker AH, Nathwani A, Thrasher AJ, Coutelle C, McKay TR, Waddington SN (2008) Luciferin detection after intranasal vector delivery is improved by intranasal rather than intraperitoneal luciferin administration. Hum Gene Ther 19 (10):1050–1056. https://doi.org/10.1089/hum.2008.023
5. Wu JC, Sundaresan G, Iyer M, Gambhir SS (2001) Noninvasive optical imaging of firefly luciferase reporter gene expression in skeletal muscles of living mice. Mol Ther 4 (4):297–306. https://doi.org/10.1006/mthe.2001.0460
6. Ryu YK, Khan S, Smith SC, Mintz CD (2014) Isoflurane impairs the capacity of astrocytes to support neuronal development in a mouse dissociated coculture model. J Neurosurg Anesthesiol 26(4):363–368. https://doi.org/10.1097/ana.0000000000000119
7. Broad KD, Hassell J, Fleiss B, Kawano G, Ezzati M, Rocha-Ferreira E, Hristova M, Bennett K, Fierens I, Burnett R, Chaban B, Alonso-Alconada D, Oliver-Taylor A, Tachsidis I, Rostami J, Gressens P, Sanders RD, Robertson NJ (2016) Isoflurane exposure induces cell death, microglial activation and modifies the expression of genes supporting neurodevelopment and cognitive function in the male newborn piglet brain. PLoS One 11 (11):e0166784. https://doi.org/10.1371/journal.pone.0166784
8. Karda R, Perocheau DP, Suff N, Ng J, Delhove J, Buckley SMK, Richards S, Counsell JR, Hagberg H, Johnson MR, McKay TR, Waddington SN (2017) Continual conscious bioluminescent imaging in freely moving somatotransgenic mice. Sci Rep 7:6374. https://doi.org/10.1038/s41598-017-06696-w
9. Karda R, Rahim AA, Suff N, Diaz JA, Perocheau DP, Martin NP, Hughes M, Delhove J, Counsell JR, Henckaerts E, McKay TR, Buckley SMK, Waddington SN (2018) Generation of light-producing somatic-transgenic mice using adeno-associated virus vector. Mol Ther 26(5):40–40
10. Gould D (2017) Mammalian synthetic promoters. Springer, New York
11. Yoder KE, Fishel R (2008) Real-time quantitative PCR and fast QPCR have similar sensitivity and accuracy with HIV cDNA late reverse transcripts and 2-LTR circles. J Virol Methods 153 (2):253–256. https://doi.org/10.1016/j.jviromet.2008.07.032
12. Chen TR, Hay RJ, Macy ML (1983) Intercellular karyotypic similarity in near-diploid cell-lines of human-tumor origins. Cancer Genet Cytogenet 10(4):351–362. https://doi.org/10.1016/0165-4608(83)90092-4
13. Barczak W, Suchorska W, Rubis B, Kulcenty K (2015) Universal real-time PCR-based assay for lentiviral titration. Mol Biotechnol 57

(2):195–200. https://doi.org/10.1007/s12033-014-9815-4

14. Kim JY, Ash RT, Ceballos-Diaz C, Levites Y, Golde TE, Smirnakis SM, Jankowsky JL (2013) Viral transduction of the neonatal brain delivers controllable genetic mosaicism for visualising and manipulating neuronal circuits in vivo. Eur J Neurosci 37 (8):1203–1220. https://doi.org/10.1111/ejn.12126
15. Werling NJ, Satkunanathan S, Thorpe R, Zhao Y (2015) Systematic comparison and validation of quantitative real-time PCR methods for the quantitation of adeno-associated viral products. Hum Gene Ther Methods 26(3):82–92. https://doi.org/10.1089/hgtb.2015.013

Chapter 12

In Vivo Bioluminescent Imaging of Marburg Virus in a Rodent Model

Shan Lei, Weijin Huang, Youchun Wang, and Qiang Liu

Abstract

Pseudoviruses are useful tools because of their safety and versatility compared to wild type viruses. Optical imaging of reporter gene labeled pseudoviruses in small animal models can allow for real-time analysis of the infection process without sacrificing the host, which has proven invaluable in the longitudinal study of disease events and testing the antiviral efficiencies of vaccine candidates, monoclonal antibodies and small molecule compounds. Here, we describe the generation of Marburg pseudovirus (pMARV) and establishment of imaging mouse model by using a deep-cooled CCD camera imager. We also describe the widespread organ distribution of pMARV during infection by ex vivo imaging of necropsied tissues. This system can significantly facilitate Marburg virus studies and enable the evaluation of treatments against MARV in BSL-2 containments.

Key words Bioluminescence, Surrogate disease model, Marburg virus, IVIS spectrum, Alternative approach

1 Introduction

Marburg virus (MARV), a member of the Filoviridae family, was first discovered in 1967, causing viral hemorrhagic fever in humans and nonhuman primates [1]. The mortality associated with MARV infection is as high as 90%, with no effective treatments, including vaccines, available to date. Though an enormous amount of research effort has been committed to the highly related Ebola virus (EBOV), there are few studies focused on MARV, and the mechanism of MARV pathogenicity remains unclear [2]. Unlike EBOV-eGFP, there is still no recombinant MARV containing a reporter gene [3]. Another big obstacle to research on MARV is the requirement for BSL-4 containments. These limitations in particular have severely hampered the progress of MARV studies and treatments.

The advent of pseudoviruses, packaging the glycoproteins of highly virulent strains to the surface of replication-incompetent

Steven Ripp (ed.), *Bioluminescent Imaging: Methods and Protocols*, Methods in Molecular Biology, vol. 2081,
https://doi.org/10.1007/978-1-4939-9940-8_12,

HIV [4], vesicular stomatitis virus [5], and feline immunodeficiency virus [6], allows researchers to safely work with the virus in BSL-2 environments. Moreover, the use of noninvasive approaches to image virus infection, without having to sacrifice animals, allows for a real-time analysis of host-virus interaction and quantitative dissemination in vivo.

Based on this, our group modified the skeleton of a pseudovirus packaging system and inserted the firefly luciferase (Fluc) gene. Then through a series of optimization processes, the package titer of HIV pseudotyped with MARV GP was greatly improved, enabling the successfully infection of mice with the MARV pseudovirus [7]. The highly predictive value of this murine model is attributable, in part, to the high sensitivity of bioluminescent imaging (BLI) technology, which has been used to study various viral pathogens by detecting optically tagged targets (luciferase or enhanced green fluorescent protein) in live animals. Therefore, imaging mouse models can be used to evaluate the antivirus efficiencies of vaccines, monoclonal antibodies and treatment drugs in vivo.

The following technique focuses on using optical imaging and subsequent data analysis to characterize MARV pseudovirus infection in a mouse model. However, similar approaches can be applied to a variety of pathogens.

2 Materials

2.1 Pseudotyped MARV Packaging

1. pcDNA3.1-MARV-GP, the MARV GP gene (Marburg virus isolate Mbg-422-2012, Uganda, GenBank: KC545387.1) was cloned into the eukaryotic expression vector pCDNA3.1.
2. pSG3.Δenv.cmv.Fluc: Cloned the Fluc gene with a CMV promoter into pSG3.Δenv at the end of the silenced *env* gene to generate pSG3.Δenv.cmv.Fluc.
3. Lipofectamine 3000. Store at 4 °C.
4. High glucose Dulbecco's modified Eagle's medium. Supplemented with 10% fetal bovine serum, and 100 U/ml penicillin and 100 μg/ml streptomycin.
5. 293T cells (ATCC, CRL-3216).
6. Incubator with a humidified atmosphere supplemented with 5% CO_2 at 37 °C.
7. Cell culture flasks.
8. 30-kDa ultrafiltration centrifugal tube.
9. 0.45 mM pore-size filter.

2.2 Titration of Pseudoviruses

1. Bright-Glo luciferase reagent: purchase from Promega, Madison, WI, USA. Store at −20 °C.
2. Glomax 96 microplate luminometer: purchase from Promega, Madison, WI, USA, or another similar microplate luminometer.
3. 96-well cell culture plates.
4. Solid black 96-well plates.
5. Sterile water.

2.3 Bioluminescence Imaging

1. Isoflurane: Store at room temperature.
2. D-luciferin: 50 mg/kg body weight. Store at −20 °C.
3. IVIS Lumina III imaging system: Purchased from Perkin Elmer, Waltham, MA, or other suitable whole animal imaging system.
4. Living Image® software (version 4): Perkin Elmer, Waltham, MA.
5. Shaver and depilate cream.

2.4 Ex Vivo Tissue Imaging

1. Sterile surgical instruments.
2. Sterile saline water.
3. Cell culture dishes.
4. Bibulous paper.

3 Methods

3.1 Pseudotyped MARV Construction and Packaging

1. Seed 293 T cells in a T150 cell culture flask. Cells should be between 80 and 90% confluent at the time of transfection.
2. Balance the Opti-MEM at room temperature.
3. Prepare different plasmids and Lipofectamine 3000 in two separate tubes as indicated in Table 1.
4. Incubate for 5 min at room temperature.
5. Combine the contents of Tube A and Tube B by adding B to A slowly. Incubate for 20 min at room temperature.
6. Add the transfection mixture to the prepared T150 cell culture flask. Do not disturb the cell monolayer.
7. Incubate for 6 h at 37 °C, 5% CO_2.
8. Remove the transfection inoculum in 6 h and add 30 ml of culture medium slowly down the side of flask without disturbing the cell monolayer (*see* **Note 1**).
9. Incubate for 42 h at 37 °C, 5% CO_2 (*see* **Note 2**).

Table 1
Lipofectamine-based transfection

Tube A	Tube B
2 ml Opti-MEM	2 ml Opti-MEM
40 μg pSG3.Δenv.cmv.Fluc	60 μl Lipofectamine 3000
20 μg pcDNA3.1-MARV-GP	
60 μl P3000	

10. Harvest the culture supernatant containing pMARV and spin at 210 × *g* for 5 min to separate cells and cell fragments from the supernatant.
11. Collect the supernatant and put it through a 0.45 mM pore-size filter to remove larger molecular substances.
12. If desired, concentrate the filtered supernatant from 30 ml to 3 ml in a 30-kDa ultrafiltration centrifugal tube (*see* **Note 3**).
13. Split the pseudovirus in 2 ml frozen pipes and store at −80 °C or liquid nitrogen until use.

3.2 Determining the Infectious Titer of the pMARV

1. Take pMARV out and dissolve in water.
2. Dilute the pMARV 50 times with complete medium (*see* **Note 4**).
3. Add 100 μl of complete medium to B2–G11 (A1–A12, H1–H12, B1–G1 and B12–G12 should be empty in this step as indicated in Table 2.
4. Add 25 μl of pMARV dilution to B2–G2 and mix well.
5. Take 25 μl from B2–G2 and serially dilute down the plate (from Column 2 to Column 10) using 8-channel pipettor.
6. Add 100 μl of 293 T cell suspension (5×10^5 cells/ml) to B2–G11.
7. Add 200 μl of sterile water to A1–A12, H1–H12, B1–G1 and B12–G12 to prevent excessive evaporation of water (*see* **Note 5**).
8. Incubate plate for 48 h at 37 °C, 5% CO_2.
9. Remove 100 μl of cell culture medium from cells.
10. Add 100 μl Bright-Glo luciferase reagent to each well and then incubate plate in dark at room temperature to react 2 min.
11. Transfer 150 μl of lysate to solid black 96-well plates for luminescence detection using Glomax 96 microplate luminometer.
12. Calculate the $TCID_{50}$ by using the Reed-Muench method.

Table 2
Serial dilutions of pMARV to determine the infectious titer

	1	2	3	4	5	6	7	8	9	10	11	12
A												
B		5^1	5^2	5^3	5^4	5^5	5^6	5^7	5^8	5^9		
C		5^1	5^2	5^3	5^4	5^5	5^6	5^7	5^8	5^9		
D		5^1	5^2	5^3	5^4	5^5	5^6	5^7	5^8	5^9		
E		5^1	5^2	5^3	5^4	5^5	5^6	5^7	5^8	5^9		
F		5^1	5^2	5^3	5^4	5^5	5^6	5^7	5^8	5^9		
G		5^1	5^2	5^3	5^4	5^5	5^6	5^7	5^8	5^9		
H												

3.3 Construction of a Bioluminescent Imaging Mouse Model

1. Prepare four different strains of mice including C57BL/6, BALB/c, KM, and NIH mice (female, 4 to 5 weeks old, four mice per group) (*see* **Note 6**).
2. Record mouse body weight.
3. Mark each of mice in the group with permanent dye colors.
4. Take pMARV out of −80 °C and dissolve in water.
5. Each mouse in four groups is infected with 0.2 ml pMARV (9.75×10^8 $TCID_{50}$/ml) via intraperitoneal (IP) injection (*see* **Note 7**).
6. Collect signals of infected mice at sequential time points (6 h, 2 day, 4 day, 5 day, 6 day, 10 day, 14 day after infection).

3.4 In Vivo Whole Animal Live Imaging

1. Animal shaving and IVIS System Preparation
   (a) All mice are shaved using a veterinary clipper to improve the sensitivity of bioluminescent detection (*see* **Note 8**).
   (b) Before imaging begins, switch on the IVIS System and open Living Image® software. Initialize the device and wait for the camera cooling down to the appropriate temperature (−90 °C as default, Fig. 1).
2. Each mouse was anesthetized by intraperitoneal (IP) injection of anesthetic solution before placement into IVIS imaging chamber (*see* **Note 9**).
3. Inject the group of mice (5 mice max per group) with substrate D-luciferin (50 mg/kg body weight) solution by IP after the mice are sedated.

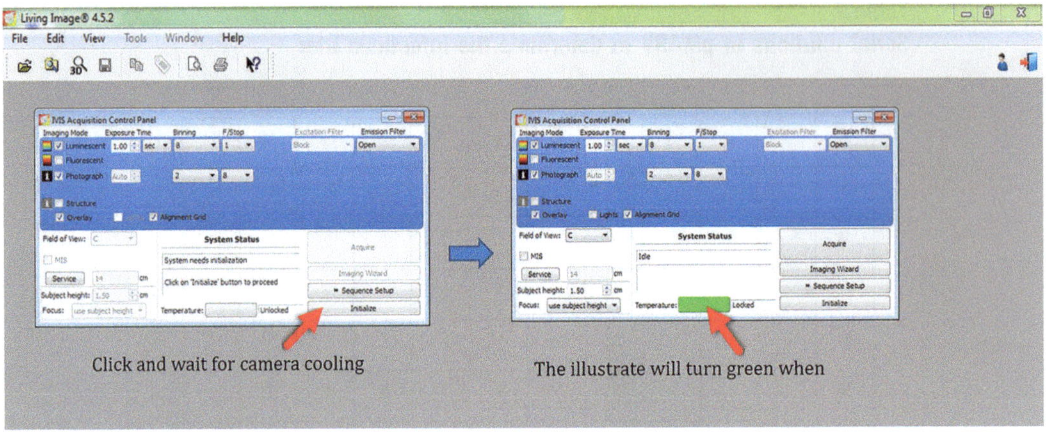

Fig. 1 IVIS system preparation

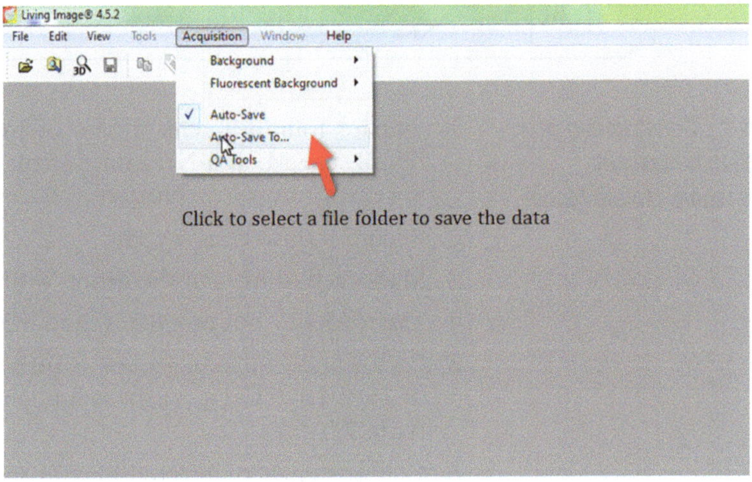

Fig. 2 The auto-save function

4. Place the group of mice ventral side up into the IVIS imaging chamber after 10 min of substrate injection and close the chamber door.
5. Between the 10 min, activate the Auto-Save function by clicking **Acquisition and select Auto-Save To**. . . on the top main function panel of the software window (Fig. 2), which will allow instant saving of all captured images. Select or create a new folder to save images (*see* **Note 10**).
6. Alter the capture parameters on the **IVIS Acquisition Control Panel** (Fig. 3), use 60 s exposure time, **C** field-of-view for 3 or fewer mice and **D** view for 5 mice.
7. (Optional) In the process of capturing, input specific information of the group in **Edit Image Labels.** The information will

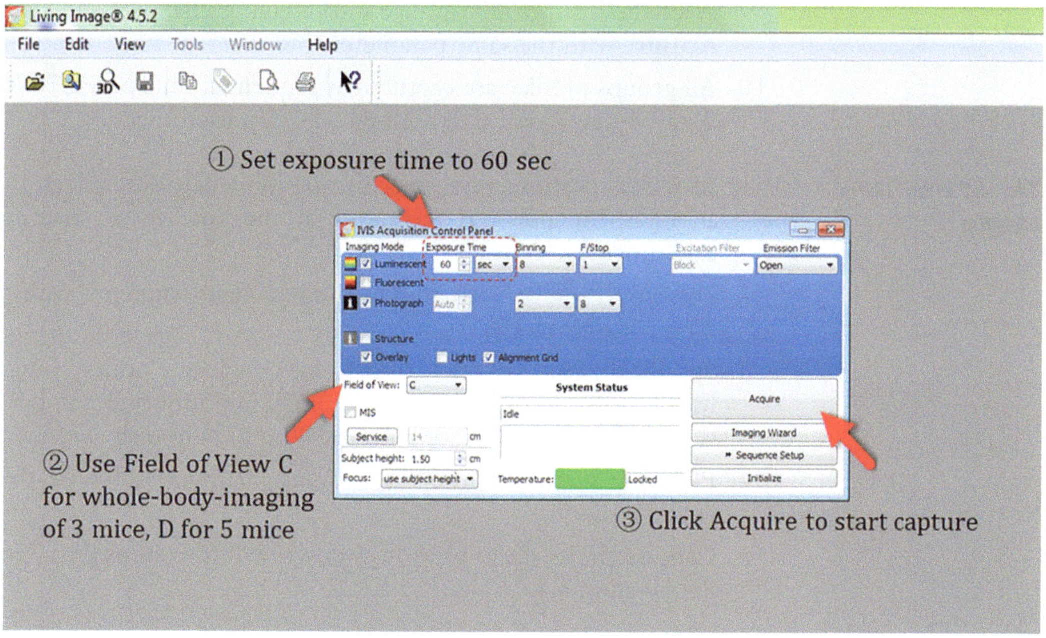

Fig. 3 Adjustments of the capture parameters

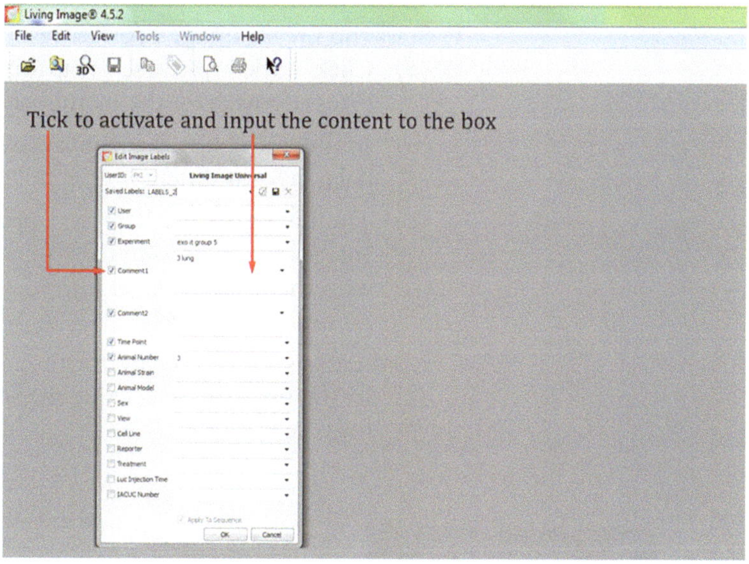

Fig. 4 Input specific information of the group in Edit Image Labels

be saved automatically with the captured data for reference (Fig. 4).

8. When imaging is completed for each group, remove animals from the imaging chamber and return animals to their cages (*see* **Note 11**).

9. Place the next group of mice into imaging chamber and click **Acquire** with the same parameters.
10. All groups of mice are captured at sequential time points (6 h, 2 day, 4 day, 5 day, 6 day, 10 day, 14 day after infection).

3.5 Ex Vivo Tissue Imaging

1. At 5 days postinfection, mice can be humanely euthanized by an AVAMA approved technique at the end of an imaging experiment.
2. Dissect the mice with sterilized surgical instruments within 3 min (*see* **Note 12**).
3. Specific organs (thymus, heart, liver, spleen, kidney, lung, lymph node, muscle skin, ovary, brain, and intestine) are harvested and carefully placed into a 100 mm petri dish (Fig. 5).
4. Place the petri dish into the imaging chamber and close the chamber door.
5. Capture the image as previously described from **steps 5** to 7 in Subheading 3.4.

3.6 Image Analysis

Once all images over the infection course have been acquired, the progression or dissemination of the infection can be analyzed with the Living Image® software.

1. Start the Living Image Browser (Fig. 6):
   (a) Click the Browse button. Alternatively, select **File→Browse** on the menu bar.
   (b) In the dialog box that appears, select the folder of sequential images at different time points and click **Select folder**.
   (c) The Living Image Browser appears. It displays all Living Image data located in the folder and its subfolders, along with the user ID, label information, and camera configuration information.
2. Load data as a group by performing one of the following:
   (a) In the browser, select the images that you want to group together.

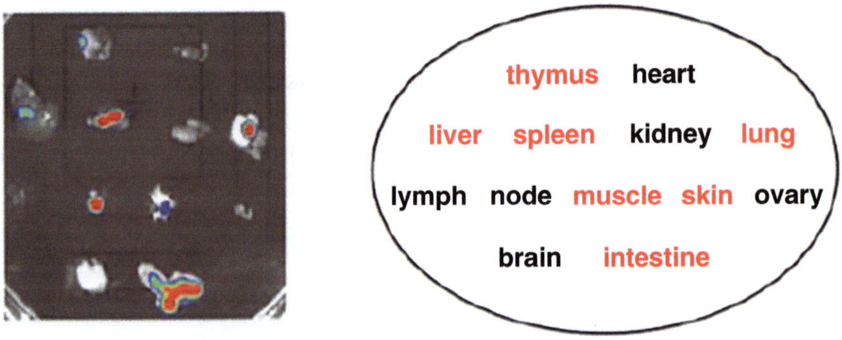

Fig. 5 Harvested organs and the placement pattern

(b) To select adjacent images in the browser, press and hold the **Shift** key while you click the first and last file in the selection.

(c) To select nonadjacent images in the browser.

(d) Press and hold the **Ctrl** key while you click the images of interest in the browser.

(e) Click **Load as Group**. The image thumbnails are displayed together in an image window (Fig. 7).

(f) Change the displayed unit to **Radiance (Photons)** in accordance with the bioluminescence imaging mode at the top-left corner of the image window (Fig. 8).

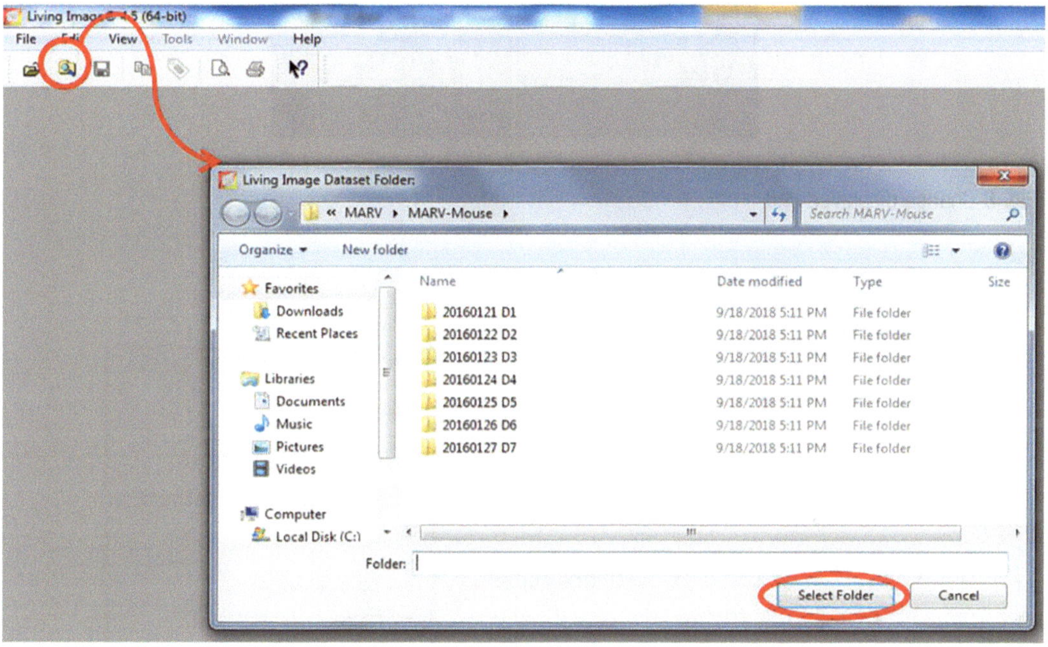

Fig. 6 Open image browser and select data folder

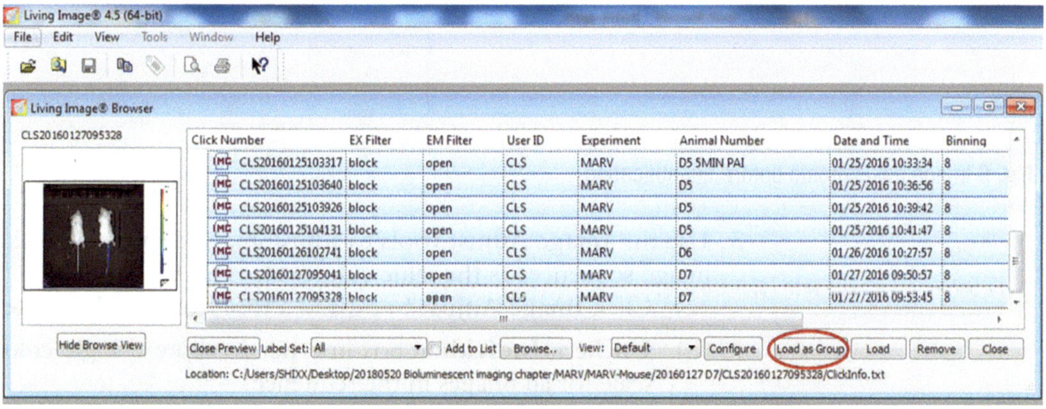

Fig. 7 Load data as group

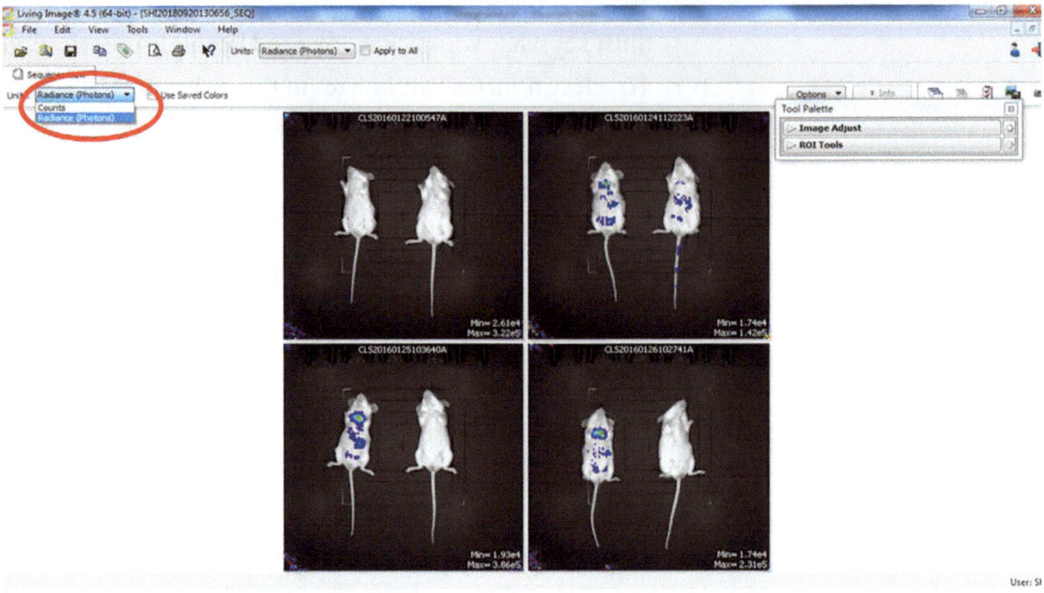

Fig. 8 Change displayed unit

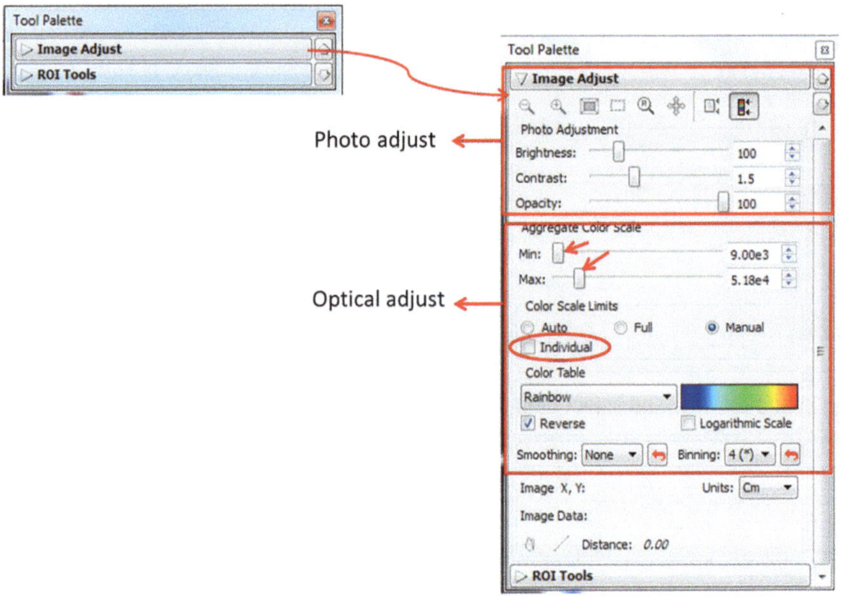

Fig. 9 Image adjustments before quantification

3. Use the **Image Adjust tools** to adjust image display and make image sequences as the same color scale (Fig. 9):
   (a) Click **Image Adjust** to expand a tool from **Tool Palette**.
   (b) Uncheck **Individual** box in order to allow a single-color scale for all images in the sequence.

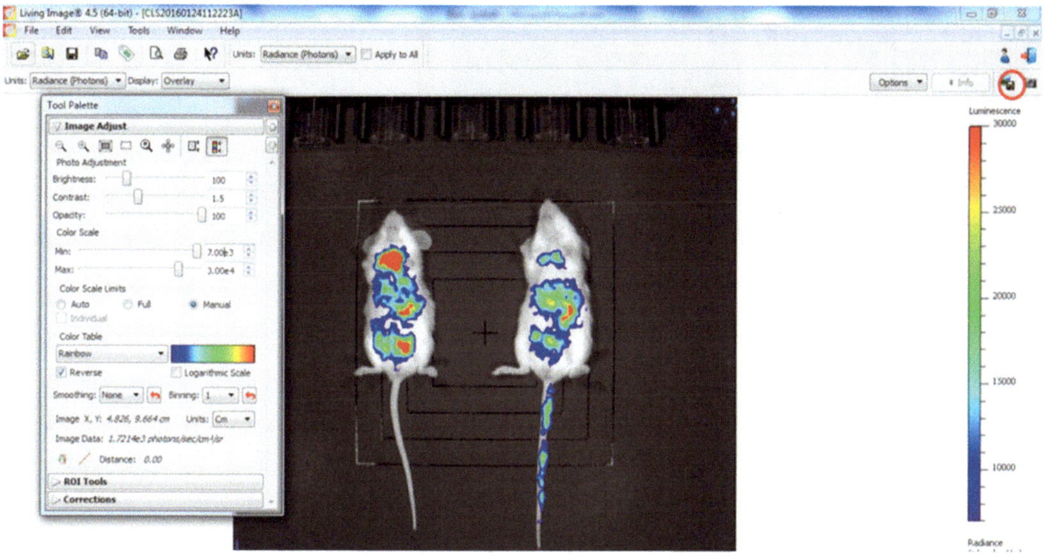

Fig. 10 Save the image as individual image

(c) Click the drop-down arrow to select color table **Rainbow** for the image data.

(d) Set the **Min** value for background subtraction. Pixels less than the color scale minimum do not appear in the image (*see* **Note 13**).

(e) Set the **Max** value for optimizing the image display. Pixels greater than the color scale maximum are displayed in the maximum color (*see* **Note 14**).

4. Save the image as individual image (Fig. 10):
   (a) Double-click each image in the sequence.
   (b) Click to save images as TIF, JPG, PNG, etc.
5. Quantitate bioluminescence to assess viral infection levels (Fig. 11):
   (a) Draw Measurement region of interest (ROI) for each animal on the Sequence.
   (b) Click to select the number of square ROIs to add to the active image.
   (c) Adjust the ROI size: Place the mouse pointer over the ROI. When the pointer becomes , click the ROI. Place the mouse pointer over an ROI handle , so that it becomes a . Drag the handle to resize the ROI. Hold Ctrl to assure all ROIs are the same size.
   (d) Adjust the ROI position: Place the mouse pointer over the ROI. When the pointer becomes a , click the ROI and drag. Click the **Measure** button.
   (e) The ROI measurements and table appear (*see* **Note 15**).

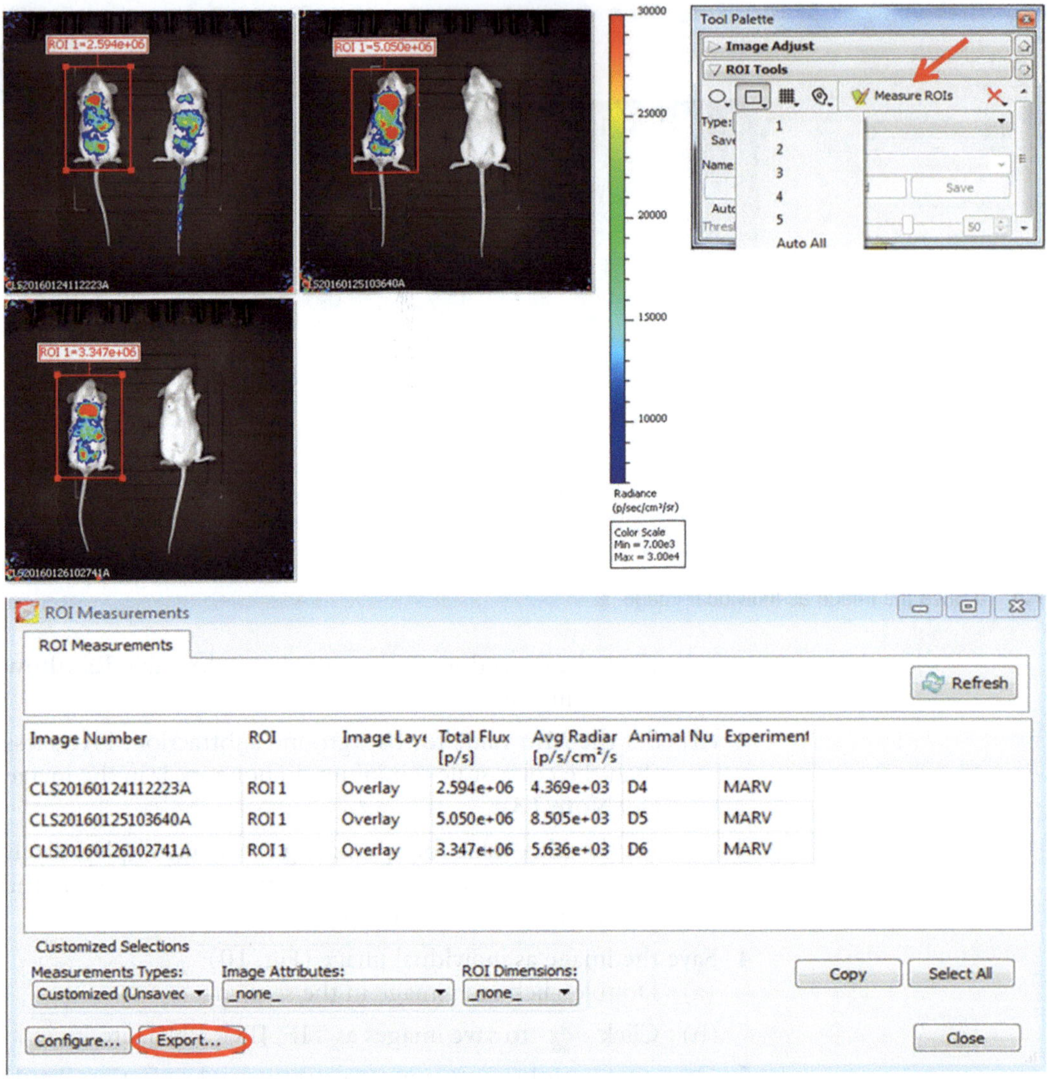

Fig. 11 Select ROI and quantify the signal

   (f) Click **Export** in the ROI Measurements table and select a file type (.txt or .csv) and click **Save.**

6. Ex vivo imaging results were analyzed as described above using **steps 1–5**.

4 Notes

1. Though it does not have to be removed from the infection inoculum in accordance to the instruction of Lipofectamine 3000, we still believe the transfection reagent has certain toxicity and would decrease the titer of pseudotyped virus based on

previous practices. The incubation time is between 4 h and 6 h, and we suggest 6 h for best results.

2. Incubation time is between 42 h and ~54 h, depending on the state of the cells.
3. With continued optimization, we have successfully built the MARV mouse model with unconcentrated pMARV. If the TCID50 is high enough, unconcentrated supernatant containing pMARV can be used directly after centrifugation and filtration.
4. If the sample is the unconcentrated supernatant, it can be used directly without dilution.
5. It can prevent the evaporation of water of the cell culture medium, so the experimental error of luminescence detection would be controlled.
6. Our previous work showed murine strains and the gender of mice could affect the susceptibility to pMARV. Of C57BL/6, BALB/c, KM, and NIH mice, better homogeneity and the highest signals were detected in BALB/C mice. Compared with female mice, male mice were more susceptive exhibiting higher signal levels, when they were infected with the same amount of pMARV. The difference between male and female mice should be considered prior to commencing the study.
7. Mice can be infected via IP and intravenous (IV) injection. We chose IP injection because of ease of handling.
8. The animal's fur will decrease sensitivity, especially with C57BL/6 mice as black fur absorbs the most signal. It is highly recommended that C57BL/6 mice are shaved prior to imaging. Though white fur will also absorb signal to a lesser extent, it doesn't have to be removed when the effect is under control.
9. Mice can be anesthetized in different ways. We used pelltobarbitalumnatricum (240 mg/kg body weight) and isoflurane. The dose of anesthetic must be carefully considered. Underdosage will lead to excessive movement and image artifacts hindering quantification. On the other hand, an overdose will be lethal.
10. The "Auto-Save" can reduce accidental loss of data.
11. It is critical to initially determine the correct luciferin kinetics postinjection for accurate and consistent quantification. The bioluminescent signal must be collected in the plateau phase typically 15 min after substrate injection. Injection route, anesthetics, body temperature and luciferin dose (mg/kg) can alter the kinetics, therefore consistency is essential.

12. Bioluminescent signal is dependent on ATP availability, which only exists in living cells. Therefore, dissected organs must be imaged immediately when removed to minimize signal loss.
13. To generate pictures for presentations or publication, the Min value can be adjusted to remove background signal, which may be 7E3, 1E4, 5E4 or 1E5, etc. This will depend on the signal-to-noise ratio.
14. The Max value can be adjusted to allow approximately 70% of the ROI to be highlighted in red. These adjustments only alter the visual appearance of the image, the radiance units detected in the ROIs are not altered.
15. We have observed different levels of autoluminescent signals in different strains of mice. It is recommended to subtract this background signal from any ROI analysis during infection.

Acknowledgments

We are grateful to Ms. Yanqiu Li, Ms. Xiaoyue Shi, and Mr. Xing Huang from PerkinElmer Inc., for technical assistance. We also would like to thank Dr. Kevin Francis, PerkinElmer Inc. and Visiting Professor at UCLA and Texas A&M, and Mr. Joshua McHattan, PerkinElmer Inc. for language editing. This study was supported by grants from the National Natural Science Foundation of China (8187131465).

References

1. Qiu X et al (2016) Two-mAb cocktail protects macaques against the Makona variant of Ebola virus. Sci Transl Med 8:329ra333
2. Cross RW et al (2018) Post-exposure treatments for Ebola and Marburg virus infections. Nat Rev Drug Discov 17:413–434
3. Wong G, Qiu X (2016) Designing efficacious vesicular stomatitis virus-vectored vaccines against Ebola virus. Methods Mol Biol 1403:245–257
4. Coffin KM et al (2018) Persistent Marburg virus infection in the testes of nonhuman primate survivors. Cell Host Microbe 24:405–416.e3
5. Mittler E et al (2018) A fluorescently labeled Marburg virus glycoprotein as a new tool to study viral transport and assembly. J Infect Dis 218:S318–S326
6. Hume A, Muhlberger E (2018) Marburg virus viral protein 35 inhibits protein kinase R activation in a cell type-specific manner. J Infect Dis 218:S403–S408
7. Zhang L et al (2017) A bioluminescent imaging mouse model for Marburg virus based on a pseudovirus system. Hum Vaccin Immunother 13:1811–1817

Chapter 13

Continuous and Real-Time In Vivo Autobioluminescent Imaging in a Mouse Model

Derek Yip, Andrew Kirkpatrick, Tingting Xu, Tom Masi, Stacy Stephenson, Steven Ripp, and Dan Close

Abstract

In vivo small animal bioluminescent imaging has become an indispensable technique for interrogating the localization, health, and functionality of implanted cells within the complex environment of a living organism. However, this task can be daunting for even the most experienced researchers because it requires multiple animal handling steps and produces differential output signal characteristics in response to a number of experimental design variables. The recent emergence of autobioluminescent cells, which autonomously and continuously produce bioluminescent output signals without external stimulation, has the potential to simplify this process, reduce variability by removing human-induced error, and improve animal welfare by reducing the number of required needlesticks per procedure. This protocol details the implantation and imaging of autobioluminescent cells within a mouse model to demonstrate how cells implanted from a single injection can be imaged repeatedly across any post-implantation timescale without the need for further human–animal interaction or signal activation steps. This approach provides a facile means to continuously monitor implanted cellular output signals in real-time for extended time periods.

Key words Autobioluminescence, Bioimaging, Drug discovery, In vivo, *lux*, Preclinical

1 Introduction

In vivo bioluminescent imaging relies upon the combination of a luciferase and its corresponding luciferin substrate to generate an optical signal. Firefly luciferase serves as the prototypical example, wherein a D-luciferin substrate reacts with ATP to form AMP, CO_2, and an oxyluciferin that releases energy in the form of photons. The inherent disadvantage of many in vivo bioluminescent imaging procedures is that the luciferin substrate must be added exogenously to initiate the bioluminescent reaction. In an animal model, this typically requires subcutaneous, intraperitoneal, or intravenous injections of the luciferin substrate prior to each desired bioluminescent measurement. The route and timing of substrate injection and its biodistribution within and clearance

Steven Ripp (ed.), *Bioluminescent Imaging: Methods and Protocols*, Methods in Molecular Biology, vol. 2081,
https://doi.org/10.1007/978-1-4939-9940-8_13,

from the animal all impact light emission kinetics [1, 2], which influences experimental outcomes and study conclusions if not carefully optimized and rigorously controlled for. Furthermore, the prerequisite for repeated luciferin injections limits the amount of imaging data that can be obtained from each animal and adversely affects animal welfare.

To circumvent the need to exogenously inject animals with a chemical substrate, autobioluminescent imaging instead uses a synthetic luciferase (derived from the *Photorhabdus luminescens luxCDABE* operon) designed specifically to express in mammalian cells using endogenous substrate scavenging for luciferin generation [3, 4]. This eliminates exogenous substrate addition, enables the continuous and autonomous emission of bioluminescence under in vitro, ex vivo, and in vivo formats, and removes the associated potential introduction of error resulting from luciferin quality variation, inconsistencies in dosage, the biological effects of foreign chemical introduction, or changes to substrate clearance rate resulting from changes in age or health among subjects [5, 6]. When paired with whole-animal imaging systems like those manufactured by PerkinElmer, MI Labs, or Spectral Instruments Imaging, autobioluminescent imaging allows small animal models to be noninvasively imaged at any point over their lifetime. This approach can therefore provide longitudinal real-time surveillance of biological processes. The techniques for working with autobioluminescent cells differ little from that of conventional bioluminescent cells expressing firefly luciferase, apart from the elimination of the multiple luciferin injection steps. The following protocol provides the details necessary for performing autobioluminescent in vivo imaging in small animal research models.

2 Materials

2.1 Autobioluminescent Cell Line

1. Autobioluminescent HEK293 human kidney cell line.

2.2 Animal Subjects

1. Five-month-old, female nu/nu (nude) immunocompromised mouse.

2.3 Cell Culture Medium

1. Dulbecco's modified Eagle's medium (DMEM).
2. Fetal bovine serum (FBS): 10% solution.
3. Penicillin/Streptomycin (PenStrep): 1% solution.
4. G418: 100 μg/mL.
5. Dulbecco's phosphate-buffered saline (DPBS).

2.4 Implantation

1. Dulbecco's phosphate-buffered saline (DPBS).
2. Trypsin: 0.05% solution.
3. Trypan Blue.
4. Isoflurane–oxygen anesthesia induction chamber.
5. Hypodermic needle: 25-gauge.
6. Syringe: 1 mL.

2.5 Hardware

1. IVIS Lumina K preclinical benchtop imaging system (PerkinElmer).
2. Hemocytometer or automated cell counting device.
3. Benchtop centrifuge capable of holding 15 mL tubes.

2.6 Software

1. Living Image 4.5.2 software (PerkinElmer).

3 Methods

3.1 Preparation of Autobioluminescent Cells

This protocol uses an autobioluminescent HEK293 human kidney cell. This cell type is robust, easy to culture, fast growing, and maintains a strong autobioluminescent output signal under a variety of conditions. However, it is important to select an appropriate cell type to support the specific goals of your experiment. If an autobioluminescent version of your cell type of choice is not readily available, the cells can be made autobioluminescent as described in **Note 1**.

1. Using a tissue culture treated T_{75} flask, culture the autobioluminescent HEK293 cell line in 15 mL of DMEM supplemented with 10% FBS, 1% PenStrep, and 100 μg/mL G418 (*see* **Notes 2** and **3**). A T_{75} culture flask should be used to ensure a sufficient number of cells are available for injection (*see* **Note 4**).
2. Incubate the cells at 37 °C in a humidified, 5% CO_2 incubator and refresh the medium every 2–3 days until the cells reach ~80% confluency.
3. When refreshing medium, the spent culture medium should be aspirated, the cells should be washed once with DPBS, then 15 mL of fresh culture medium should be added.
4. Upon reaching 80% confluence, the cells are ready to be harvested and prepared for implantation as described below.

3.2 Implantation of Autobioluminescent Cells into the Mouse Subject

This protocol details the subcutaneous implantation of autobioluminescent cells into the dorsal flank of an immunocompromised mouse subject. However, alternative injection sites such as tail vein, intraperitoneal, or intraorganellar; or an alternative subject type can be used based on the ultimate objective of the study (*see* **Note 5**). All animal studies should be conducted in compliance with protocols approved by an appropriate institutional animal care and use committee and the highest ethical standards should be employed. This protocol was conducted in accordance and compliance with all relevant regulatory and institutional agencies, regulations and guidelines.

1. Harvest the prepared autobioluminescent cells by aspirating the culture medium, washing once with 5 mL of 1× DPBS, and disassociating with 3 mL of 0.05% trypsin.
2. Incubate the cells in trypsin for ~2 min, or until they have detached, at 37 °C in a humidified, 5% CO_2 incubator.
3. Once the cells have detached, recover them into a 15 mL tube by adding 7 mL of DMEM supplemented with 10% FBS and 1% PenStrep to bring the total volume to 10 mL and transferring the full volume to the 15 mL tube.
4. Count the cells using a hemocytometer or automated cell counting system (*see* **Note 6**).
5. Concurrent with the cell counting procedure, centrifuge the remainder of the harvested cells at 250 × *g* for 5 min.
6. Aspirate the supernatant from the centrifuged cells and resuspend at a concentration of 1×10^7 cells/mL in DPBS (*see* **Note 7**).
7. Prior to implantation, anesthetize the mouse subject to be assayed by placing it into an anesthesia induction chamber and supplying 5% (v/v) inhaled isoflurane in 1 L/min of oxygen. Closely monitor the status of the subject inside the induction chamber. The isoflurane level may need to be adjusted accordingly to achieve desired anesthesia level. Be careful to not overanesthetize the subject.
8. Using a 25-gauge needle and a 1 mL syringe, draw up the resuspended cells from **step 6** into the syringe and inject 100 μL subcutaneously into the dorsal flank (*see* **Note 8**).

3.3 Imaging Autobioluminescent Output Signals

This protocol details the imaging of a single mouse subject in the IVIS Lumina K imaging system over a period of 30 min. Depending on the brand or model of imaging equipment being used, it may be possible to image multiple subjects simultaneously. Fortunately, if this is the case the same steps can be performed in parallel on each individual subject. Alternatively, the steps can be iterated across multiple subjects in series to achieve the same result.

1. Using the Living Image software, initialize the IVIS Lumina K and set the stage temperature to 37 °C.

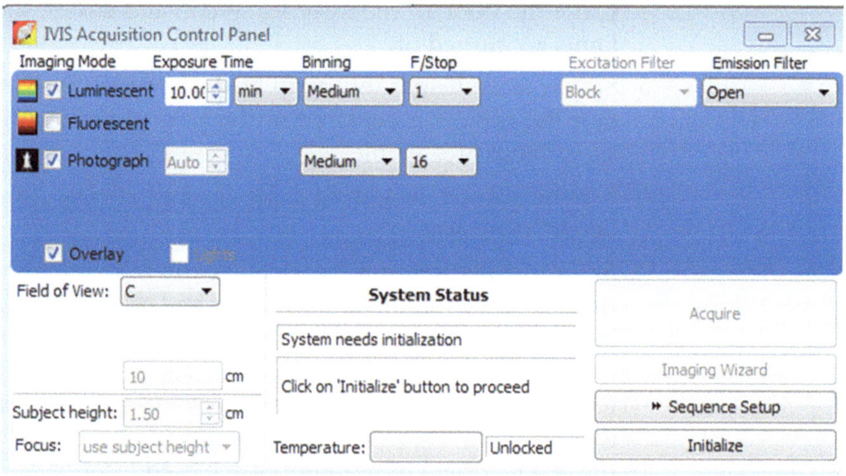

Fig. 1 An example of the image acquisition settings for capture of autobioluminescent signal from subcutaneously implanted cells. Some of these settings, such as the binning, can be adjusted following image acquisition, but most cannot. It is important to ensure that the Photograph and Overlay checkboxes are selected in addition to the Luminescent checkbox when acquiring bioluminescent output signals. Depending on the autobioluminescent cell type used, the number of cells implanted, and the health of the implanted cells, the exposure time can be reduced down to as little as 1 s while still acquiring sufficient flux for pseudocolor localization

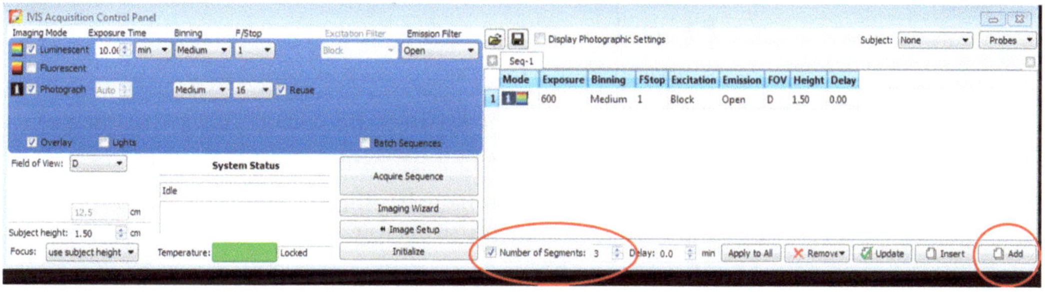

Fig. 2 Example of an image capture sequence that will interrogate the subject continuously for half an hour using three 10-min exposures. The Add button and the check box for adjusting the number of segments in the sequence are circled

2. Manually adjust the in-chamber anesthesia system to deliver 1.5% (v/v) inhaled isoflurane in 1 L/min of oxygen (*see* **Note 9**).
3. Place the anesthetized subject harboring the injected autobioluminescent cells into the imaging chamber of the IVIS Lumina K. Be sure to position the subject so that the area harboring the injected cells is facing toward the camera and so that the subject's nose fits within the nose cone. Plug any unused nose cones with rubber stoppers to prevent the loss of anesthesia.

4. Close the door to the imaging system and ensure the interior lights are turned off.
5. Set the stage height setting to field of view C (*see* **Note 10**) so that the subject is centered in the field of view with minimal superfluous area visible.
6. Set the subject height to 1.50 cm and choose "use subject height" for focus.
7. Acquire a photograph by selecting the "Photograph" setting from the list of illumination settings and clicking the "Acquire" button. This photograph will show the position of the subject and allow you to determine if any changes in positioning need to be made prior to autobioluminescent signal acquisition. If repositioning is required, this step should be repeated to ensure the proper orientation has been achieved to provide an appropriate final image.
8. Once the subject has been appropriately positioned, select the "Luminescent" setting (both the "Photograph" and "Luminescent" settings should now be checked) from the list of illumination settings and input the following imaging acquisition settings (Fig. 1, *see* **Note 11**):
   Exposure Time: 10 min.

   Binning: Medium.

   F Stop: 1.

   Emission Filter: Open.
9. After confirming the settings are correct, click the "Sequence Set Up" button to open the sequence editor dialog.
10. Click the "Add" button to import your acquisition settings into the sequence editor dialog.
11. Check the "Number of Segments" box and adjust the corresponding number to 3. This will take three 10 min exposure images for a total of 30 continuous minutes of observation (Fig. 2).
12. After the imaging process completes, it can be repeated to obtain further data, or the subject can be returned to its cage and monitored until fully recovered from anesthetic. This recovery usually occurs within 5 min (*see* **Note 12**).

3.4 Processing Autobioluminescent Output Data

This protocol details the interrogation of a single implantation location from a single subject at one time point. The steps listed in this section can be duplicated for additional regions of interest, subjects, or time points as needed.

1. Double click the first image in the series to open that image for processing (Fig. 3).

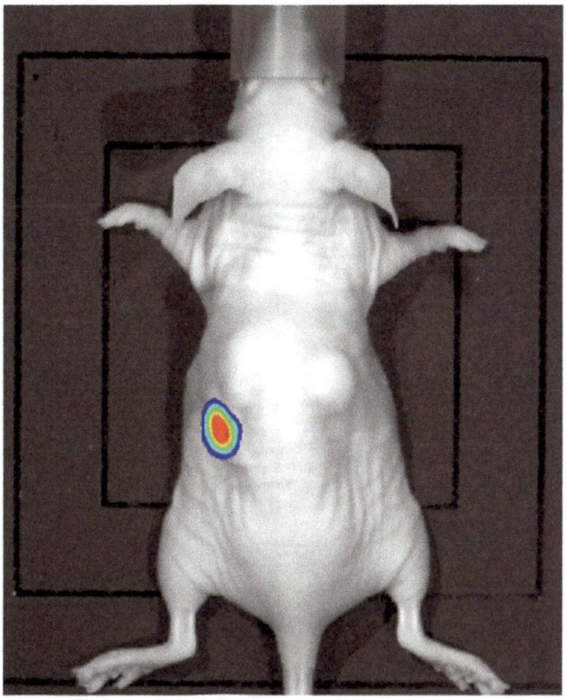

Fig. 3 Example of a pseudocolor image showing the location of subcutaneously implanted autobioluminescent cells within a nude mouse model. The location of the injected cells should be easily visible from the autobioluminescent signal

2. Open the "ROI Tools" section of the Tool Palette in the Living Image software.
3. Select the circular ROI tool from the menu and use the cursor to draw a circle around the region of interest (ROI) in the pseudocolor image (Fig. 4, *see* **Notes 13** and **14**).
4. After the ROI has been drawn, click the "Measure ROI's" button and a dialog box will open displaying the output signal level within the ROI (Fig. 5, *see* **Note 15**).
5. Click the "Select All" button to select the displayed data.
6. Click the "Copy" button to copy the information to the clipboard.
7. Paste the information into Microsoft Excel, or your preferred program for data analysis and storage.
8. Amend the file with any pertinent metadata that was not available for direct import from Living Image, save the file, and proceed with data analysis as appropriate.

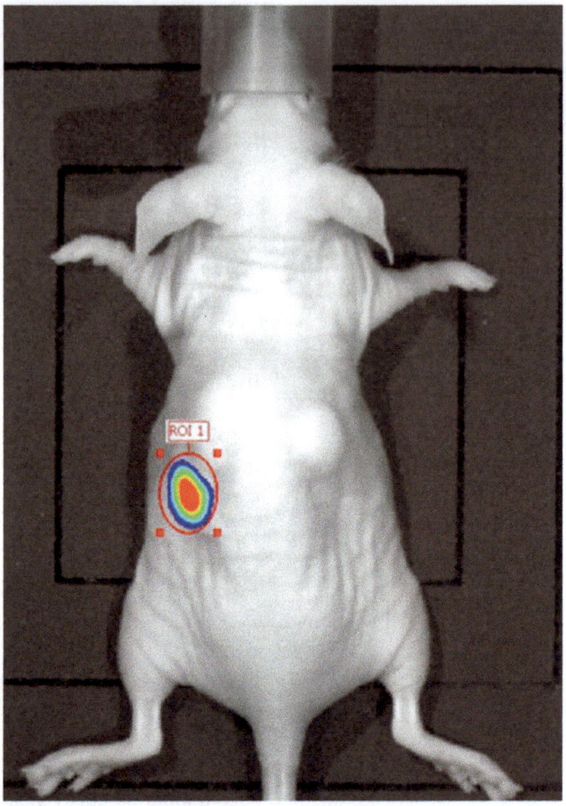

Fig. 4 A region of interest (ROI) drawn around the location of the implanted autobioluminescent cells allows total signal output to be quantified

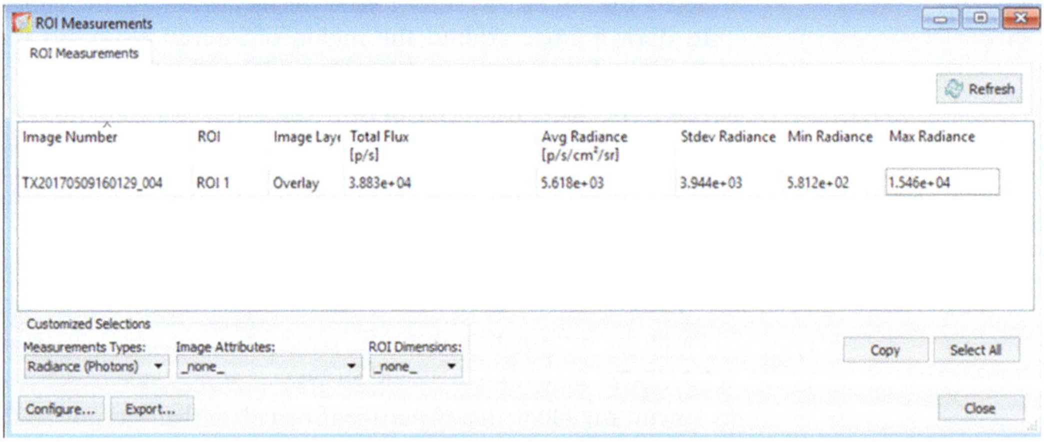

Fig. 5 An example of the ROI measurement dialog box. These data show the quantification of autobioluminescent signal from within the ROI and can be used to make comparisons between different locations, time points, or treatments. Note the Configure button in the bottom left of the box. This will allow you to select additional metadata beyond the minimal default data (shown)

4 Notes

1. If an autobioluminescent version of the necessary cell type for your experimental design is not readily available, the autobioluminescent phenotype can be genetically encoded to your cell line of interest via transient or stable transfection as described in [7].
2. The inclusion of 100 μg/mL G418 will ensure maintenance of the genetically encoded autobioluminescent phenotype. Depending on the source of the autobioluminescent cell line, this level of selective pressure may need to be increased or decreased. Always follow the supplier's recommended culture conditions or determine the ideal selective pressure empirically.
3. If a cell type other than HEK293 is used, it may be necessary to use a different medium formulation.
4. A T_{75} flask provides a 75 cm^2 culture area. Alternative culture formats can be used to scale up or down, or multiple flasks can be used to obtain additional cells without necessitating deviation from the volumes listed in this protocol.
5. Subcutaneous implantation deposits the autobioluminescent cells relatively near the surface of the subject, which limits the effects of absorption and dispersion on the bioluminescent output signal. If performing deep tissue implantation, it may be necessary to increase the number of implanted cells or the signal acquisition time to achieve similar levels of detection.
6. It is helpful to preprepare any necessary supplies for counting the cells before beginning the harvesting procedure. Most often, this entails the preparation of a 10 μL aliquot of Trypan Blue into a 1.5 mL microcentrifuge tube. Doing so will allow for 10 μL of the cell resuspension to be rapidly transferred to this preprepared aliquot and mixed 1:1 for live/dead counting.
7. Resuspension of the cells at a concentration of 1×10^7 cells/mL will provide for 1×10^6 cells to be implanted in a 100 μL volume. This number can be adjusted based on the desired delivery volume or total number of implanted cells that are needed for a particular experimental design but represents a good starting point for evaluating the performance of most autobioluminescent cell lines.
8. Care should be taken to ensure the injection is not made too deep. It is highly recommended that the practitioner work under the supervision of an experienced technician until they are confident in their ability to handle the anesthetized subject and perform the injection efficiently.

9. Slight adjustment of the isoflurane level may be necessary to keep the subject anesthetized throughout the entire imaging period depending on subject weight and system calibration.
10. There are four fields of view ranging from A (the closest in length to the camera) to D (the farthest). It may be necessary to select an alternate field of view if using multiple subjects, if an alternative subject type is used, or depending on the size and orientation of the subject being used.
11. These settings are optimized for the detection of low levels of autobioluminescent production and work well for acquiring autobioluminescent output signals from most applications without necessitating more total time per image than would be required for substrate-requiring luciferase reporter systems. It is often possible to reduce the exposure time significantly below the listed 10 min interval, which can reduce the overall time required per image.
12. Because the autobioluminescent signal is continuous and does not exhaust a finite supply of externally supplied luciferin like bioluminescent reporter systems do, the subject can be imaged at any desired timepoint, or as often as needed for repeated image acquisition, without the introduction of error due to changes in the output signal strength of the autobioluminescent cells.
13. The circular ROI tool is commonly used because it often fits neatly around the mass of implanted cells. However, any of the alternative ROI drawing tools may be used if they are deemed more appropriate.
14. To apply this same ROI at the same position to all images within the sequence, right click the ROI and select "Copy ROI," then right click and select "Paste ROI" in each additional image. This will allow you to make direct comparisons of the same region among all acquired images.
15. Note that clicking the "Configure" button in this dialog box will allow you to include image metadata detailing a wealth of different data points. It is advisable to review these options prior to completing the next steps to ensure you are capturing sufficient metadata for downstream analysis.

Acknowledgments

Research support was provided by the US National Institutes of Health under award numbers NIMH-1R43MH118186, NIGMS-1R43GM112241, NIGMS-1R41GM116622, NIEHS-2R44ES022567, and NIEHS-1R43ES026269 and the US National Science Foundation under award number CBET-1530953.

References

1. Inoue Y, Kiryu S, Izawa K, Watanabe M, Tojo A, Ohtomo K (2009) Comparison of subcutaneous and intraperitoneal injection of D-luciferin for *in vivo* bioluminescence imaging. Eur J Nucl Med Mol Imaging 36(5):771–779
2. Lee K, Byun S, Paik J, Lee S, Song S, Choe Y, Kim B (2003) Cell uptake and tissue distribution of radioiodine labelled D-luciferin: implications for luciferase based gene imaging. Nucl Med Commun 24(9):1003
3. Close D, Xu T, Smartt A, Rogers A, Crossley R, Price S, Ripp S, Sayler G (2012) The evolution of the bacterial luciferase gene cassette (*lux*) as a real-time bioreporter. Sensors 12(1):732–752. https://doi.org/10.3390/s120100732
4. Xu T, Ripp S, Sayler G, Close D (2014) Expression of a humanized viral 2A-mediated *lux* operon efficiently generates autonomous bioluminescence in human cells. PLoS One 9(5): e96347. https://doi.org/10.1371/journal.pone.0096347
5. Class B, Thorne N, Aguisanda F, Southall N, McKew JC, Zheng W (2015) High-throughput viability assay using an autonomously bioluminescent cell line with a bacterial *lux* reporter. J Lab Autom 20(2):164–174. https://doi.org/10.1177/2211068214560608
6. Xu T, Kirkpatrick A, Toperzer J, Ripp S, Close D (2019) Improving estrogenic compound screening efficiency by using self-modulating, continuously bioluminescent human cell bioreporters expressing a synthetic luciferase. Toxicol Sci 168(2):551–560. https://doi.org/10.1093/toxsci/kfz004
7. Close D, Xu T, Ripp S, Sayler G (2014) Real-time bioluminescent tracking of cellular population dynamics. In: Badr C (ed) Bioluminescent imaging: methods and protocols. Humana Press, New York, NY, pp 107–116

Chapter 14

In Vivo Tracking of Tumor-Derived Bioluminescent Extracellular Vesicles in Mice

Prakash Gangadaran and Byeong-Cheol Ahn

Abstract

Extracellular vesicles (EVs) are nano-membrane vesicles containing exosomes and microvesicles, and are released by almost all types of cells. They can carry lipids, proteins, mRNAs, and miRNAs to enable intercellular communication between cells either locally or distantly without direct cell-to-cell contact. Cancer-derived EVs are known to facilitate tumor progression and metastasis by preparing premetastatic niches. Here, we define a strategy to label cancer derived EVs with *Renilla* luciferase for noninvasive bioluminescence imaging (BLI) and monitoring of intravenously administered EVs in vivo.

Key words Extracellular vesicles, Cancer cells, Bioluminescence, Renilla luciferase, Coelenterazine

1 Introduction

Extracellular vesicles (EVs) are membrane nano-sized vesicles produced by cells that are released into the cell culture media and intracellular space and are detectable in body fluids [1]. EVs range from 50 to 400 nm in diameter and are capable of transporting lipids, DNAs, RNAs, and proteins to recipient cells [2]. These vesicles include exosomes (50–200 nm) and microvesicles (50–400 nm). EVs can also be internalized into cells so they can mediate intercellular communication between neighboring and distant cells [3]. Importantly, EVs have been used as nanocarriers for drugs and as endogenous therapeutic agents for cancers and various ischemic diseases [2, 4, 5].

Cancer cells are known to produce large quantities of EVs as compared to normal cells [6]. Cancer cell-derived EVs contain a select group of abundant proteins and enriched amounts of nucleic acids (such as mRNAs, miRNAs, and DNA). Such an extreme character of cancer cell-derived EVs modulates normal cells in their tumor microenvironment to promote tumor progression [7, 8]. It is therefore necessary to elucidate the visualization and monitoring of intravenously administered cancer derived-EVs in vivo.

Steven Ripp (ed.), *Bioluminescent Imaging: Methods and Protocols*, Methods in Molecular Biology, vol. 2081,
https://doi.org/10.1007/978-1-4939-9940-8_14, © Springer Science+Business Media, LLC, part of Springer Nature 2020

Bioluminescent imaging using luciferase expression can reliably visualize cells in living animals and importantly the advantage of bioluminescent imaging is that it allows detection of very low levels of photon signals from animals due to the absence of background signals [9]. *Renilla* luciferase (Rluc) is a blue-light-emitting, single-subunit and cofactor-less luciferase derived from the marine anthozoan *Renilla reniformis* (Pallas, 1766), a sea pansy species of the colonial cnidarian in the family Renillidae [10]. Rluc catalyzes oxidative decarboxylation by using molecular oxygen. Relaxation of the electronically excited coelenteramide reaction product is accompanied by emission of a photon of blue light (~470 nm) [11–13]. This makes Rluc an ideal reporter for the visualization and monitoring of EVs in vivo.

For the EV labeling of cancer cells, a human thyroid cancer cell line (Cal62) was used. By overexpressing Rluc in Cal62 cells followed by EV isolation, we could successfully visualize and monitor the systemically administered EVs in nude mice through in vivo bioluminescence imaging.

2 Materials

2.1 EV Reporter Construct and Cancer Cells

1. EV reporter construct: pCMV-Rluc-IRES-Puro^r, a lentivirus vector encoding Rluc and puromycin resistant gene, separated by an internal ribosomal entry site (IRES). All under the control of the constitutively active cytomegalovirus (CMV) promoter.
2. Human anaplastic thyroid cancer cell (Cal62) (American Type Culture Collection, ATCC; Manassas, VA, USA).
3. Dulbecco's Modified Essential Medium (DMEM), supplemented with 10% fetal bovine serum (FBS) and 1% penicillin–streptomycin.
4. Tissue culture plates, 6-well and 24-well.
5. Syringe Filter: 0.20 μm membrane filter.
6. 10 ml syringes.
7. EV-depleted FBS: FBS was passed through a 0.20 μm syringe filter and ultra-centrifuged for 18 h at 120,000 × g at 4 °C. Collect supernatant (EV-depleted FBS). Aliquot EV-depleted FBS into 50 ml conical tubes and store at −80 °C.
8. EV-isolation medium: DMEM supplemented with 10% of FBS (EV-depleted FBS) and 1% penicillin–streptomycin.

2.2 EV Isolation Components

1. Ultracentrifuge: For example, Beckman-Coulter Optima™ L-100 XP.
2. Rotor: For example, Beckman-Coulter Type SW28 rotor.

3. Ultracentrifuge tubes: Ultra-Clear™ tubes 1 × $3^1/_2$ (25 × 89 mm) (Beckman-Coulter, CA, USA).
4. Syringe filter: 0.45 μm and 0.20 μm membrane filters.
5. 10 ml disposable plastic syringes.
6. Double filtered phosphate buffered saline (PBS): Prepare PBS using PBS tablets (137 mM sodium chloride, 2.7 mM potassium chloride and 10 mM phosphate buffer) and filter twice through a 0.20 μm syringe filter.

2.3 Rluc Activity Assay

1. An optical imaging system such as the IVIS Lumina II In Vivo Imaging System (Perkin-Elmer, Waltham, MA, USA)
2. 96-well black microtiter plates, clear bottom with lid.
3. Coelenterazine (CTZ), the Rluc substrate: RediJect Coelenterazine H (ready-to-use) (PerkinElmer, USA).

2.4 In Vivo Bioluminescence Imaging

1. Animal: BALB/c nude mice or any other mouse of choice.
2. Isoflurane.
3. IVIS Lumina II In Vivo Imaging System with XGI-8 Gas Anesthesia System (Perkin-Elmer, Waltham, MA, USA).
4. Double filtered PBS (see Subheading 2.2.6).
5. 1.5 ml brown microcentrifuge tubes.
6. Coelenterazine: prepare fresh CTZ before each injection [20 μl of RediJect Coelenterazine H (ready-to-use) and 180 μl of sodium phosphate (pH -8.5) per mouse)]. Final concentration of 15 μg/mL.
7. Insulin syringe needles: 31 Gauge, 1 ml, 5/16″ needle (8 mm).

3 Methods

3.1 Generate EV Donor Cells

1. Seed 0.1 × 10^6 HEK293T cells per well in fresh culturing medium in a 24-well tissue culture plate (growth area 1.9 cm^2).
2. When 50–60% confluence is reached, transduce the cells with lentiviral vectors encoding Rluc-IRES-$Puro^R$ in 0.5 ml medium supplemented with 2 μl polybrene (5 mg/ml) for each well.
3. 48 h post-transduction examine Rluc reporter expression under an IVIS Lumina II to determine transduction. The optimal lowest concentration (kill curve) of puromycin to kill 100% of un-transfected cells is determined by incubation of Cal62 cells, plated in 6-well plates, with various concentrations of puromycin ranging from 1 to 10 ng/ml. (We used 6 ng/ml puromycin for selection of stable clones). Grow the transfected cells in this medium for 2 weeks before selection of stable clones with puromycin. Cells were used to generate EVs labeled with Rluc.

3.2 EV Isolation and Rluc Activity Assay

1. Add 2×10^6 of Cal62 cells expressing Rluc per 100 mm plate in fresh medium.
2. When the cells reach approximately 50% confluence, gently aspirate the growth medium, wash the cells with PBS in the plates (twice) and replace with 10 ml of fresh EV-depleted medium.
3. Incubate the cells for another 48–70 h at 37 °C, 5% CO_2 to generate conditioned medium containing EVs.
4. When cells reach 90% confluence, collect conditioned medium in 50 ml conical tubes and centrifuge at $300 \times g$ for 10 min at 4 °C to remove live cells. (*see* **Note 1**).
5. Collect supernatant in separate 50 ml conical tubes and then centrifuge at $1500 \times g$ for 20 min to remove cell debris.
6. Collect supernatant in separate 50 ml conical tubes and then centrifuge at $2500 \times g$ for 20 min to remove apoptotic bodies.
7. Prepare a 10 ml syringe attached to a 0.45 μm filter.
8. Place the assembled filtering syringe into a 50 ml conical tube and filter the supernatant through the syringe. (*see* **Note 2**).
9. Transfer the filtered supernatant into Ultra-Clear™ tubes (35 ml/tubes) and move to centrifuge buckets (Fig. 1a1–a3) (*see* **Note 3**).

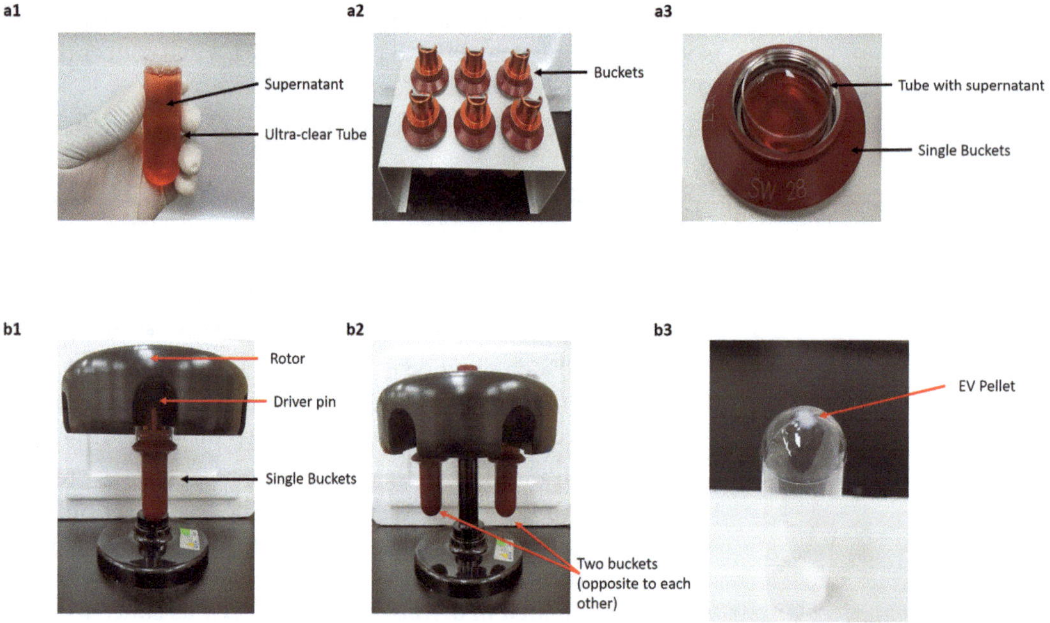

Fig. 1 EV isolation procedure. (**a**) Placement of supernatant into centrifuge buckets: (**a1**) Conditioned medium collection into ultracentrifuge tubes (35 ml/tubes). (**a2**) Buckets of six shown in the stand used for ultracentrifugation. (**a3**) Place tubes into buckets. (**b**) Ultracentrifugation setup and isolation of EVs: (**b1**) Hang the buckets on driver pins in the rotor, (**b2**) Hang the buckets opposite to one another in the rotor for balance. 2, 4, or 6 buckets can be used according to volume (ml) of supernatant. (**b3**) Location of EV pellet after aspiration of supernatant

10. In a Type SW28 rotor, place tubes diagonally opposite to one another for balance and centrifuge at 100,000 × g for 60 min at 4 °C. (Fig. 1b1, b2)
11. Carefully remove tubes from the rotor with minimal agitation. Locate the EV pellet at the bottom of the tubes. (*see* **Note 4**).
12. Gently aspirate the supernatant without agitation using a Pasteur pipette, making sure pellets are not disturbed. (*see* **Note 5**).
13. Add 1 ml of PBS and mix gently with 1 ml pipette and then add 37 ml of PBS to tubes
14. Centrifuge at 100,000 × g for 60 min at 4 °C.
15. Carefully remove tubes from the rotor with minimal agitation. Locate the EV pellet at the bottom of the tubes.
16. Place the tubes upside down on to filter paper for 2–3 min to allow remaining supernatant to be removed from the walls of the centrifuge tubes. The pellet is readily visible at this stage. (Fig. 1b3) (*see* **Note 6**).
17. Add 100 μl PBS to resuspend EV pellets by pipetting up and down and transfer to 1.5 ml microcentrifuge tube.
18. Determine protein concentration of the EV sample using preferred protein quantification methods.
19. Use immediately or store the samples at −80 °C.

3.3 Rluc Assay on Isolated EVs

1. Dilute CTZ in PBS and protect from light.
2. Dilute isolated EVs in PBS at 1:100 ratio. Add 100 μl of diluted EVs per well in triplicates into a 96-well black microtiter plate, clear bottom with lid.
3. Add 2 μl of CTZ (10 μg/mL final concentration) to each well and acquire image using IVIS Lumina II In Vivo Imaging System.
4. An increase in signal (100,000 or more-fold) of EV-Rluc derived from Cal62/Rluc cells, when compared to negative or blank controls, confirms the successful labeling of Rluc to EVs.

3.4 In Vivo EV Bioluminescence Imaging

1. To visualize EV-Rluc in mice in vivo, prepare 25 μg of EV-Rluc in a volume of 100 μl (adjusted with PBS) in a 1.5 ml microcentrifuge tube.
2. Anesthetize the animal by isoflurane (2.5%) using the XGI-8 Anesthesia System and inject the EV-Rluc (25 μg/mice) into the tail vein of the mouse using an insulin syringe needle.
3. For each mouse, prepare 20 μl of RediJect Coelenterazine H (ready-to-use) and 180 μl of sodium phosphate in a 1.5 ml microcentrifuge tube. Mix well with pipette before injection.

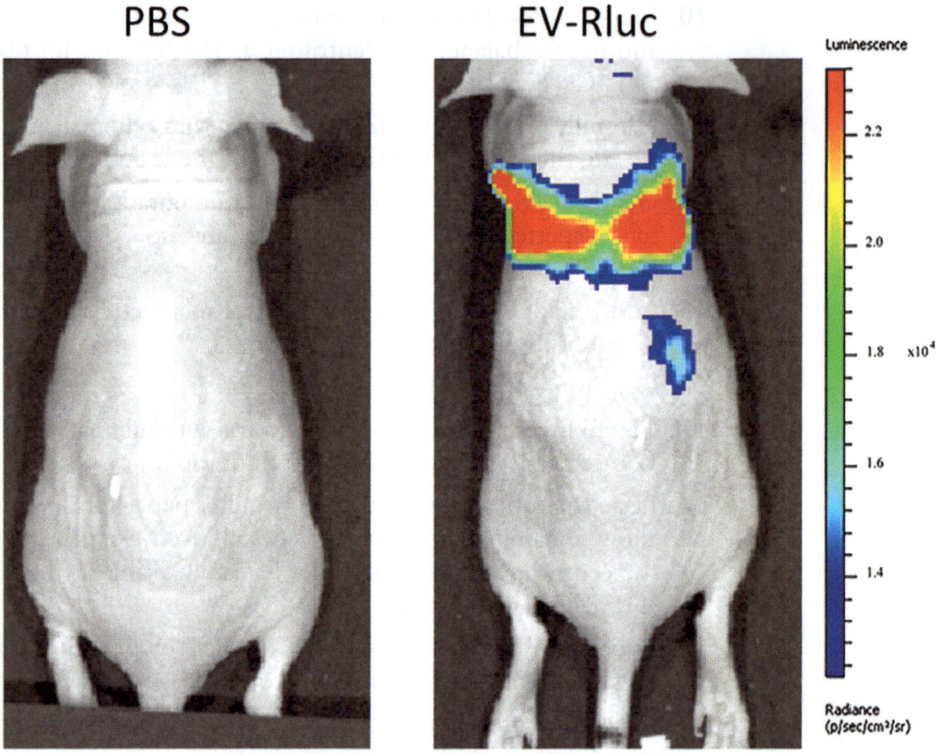

Fig. 2 In vivo bioluminescence imaging of cancer cell (Cal62/Rluc)-derived EVs. Bioluminescent image of nude mice administered with either PBS (control) or EV-Rluc. Bioluminescence signals were demonstrated in both lung fields at 10 min after intravenous injection of EV-Rluc

4. Ten minutes following EV injection, anesthetize the animal by isoflurane (2.5%) using the XGI-8 Anesthesia System.
5. Inject 200 μl of CTZ mixture through the tail vein with an insulin syringe needle.
6. Immediately following CTZ administration, image the animal for bioluminescence signal using the IVIS Lumina II In Vivo Imaging System (Fig. 2).

4 Notes

1. Avoid cells growing to more than 90% confluence, as overgrowth of cells within the plate area can result in cell death, which leads to production of apoptotic bodies. Small size (less than 300 nm) bodies may end up with EVs population.
2. The filter and filtering syringe can be reused between filtrations. Avoid pulling the plunger with filter on, as this will help to maintain the condition of filter membrane. When the syringe

becomes difficult to press, replace the filter to avoid breakage of filter and squashing the larger vesicles to pass though the filters.

3. Always use Ultra-Clear™ tubes to spot the EV pellets, which help to retain all of the isolated EVs. Cancer cell-derived EVs are always visible in Ultra-Clear™ tubes since cancer cells produce large quantities of EVs that are visible in the clear tubes.
4. EV pellets detach easily from the tube bottom when the tubes undergo sudden movement, which may lead to loss of EVs into supernatant.
5. Remove the 30 ml supernatant using an aspiration pipette and try to remove remaining supernatant as much as possible using 1 ml hand pipette.
6. Keep the tubes upside down to remove the residual supernatant in the wall as it will result in a significant increase in protein concentration, which will affect the absolute concentration of EVs and Rluc signal in subsequent assays.

Acknowledgments

This research was supported by a grant of the Korea Health Technology R&D Project through the Korea Health Industry Development Institute (KHIDI), funded by the Ministry of Health & Welfare, Republic of Korea (grant number: HI15C0001).

References

1. Gangadaran P, Hong CM, Ahn B-C (2018) An update on in vivo imaging of extracellular vesicles as drug delivery vehicles. Front Pharmacol 9:169. https://doi.org/10.3389/fphar.2018.00169
2. Gangadaran P, Rajendran RL, Lee HW et al (2017) Extracellular vesicles from mesenchymal stem cells activates VEGF receptors and accelerates recovery of hindlimb ischemia. J Control Release Off J Control Release Soc 264:112–126. https://doi.org/10.1016/j.jconrel.2017.08.022
3. Gangadaran P, Hong CM, Ahn B-C (2017) Current perspectives on in vivo noninvasive tracking of extracellular vesicles with molecular imaging. Biomed Res Int 2017:9158319. https://doi.org/10.1155/2017/9158319
4. Zhu L, Oh JM, Gangadaran P et al (2018) Targeting and therapy of glioblastoma in a mouse model using exosomes derived from natural killer cells. Front Immunol 9:824. https://doi.org/10.3389/fimmu.2018.00824
5. Zhu L, Gangadaran P, Kalimuthu S et al (2018) Novel alternatives to extracellular vesicle-based immunotherapy - exosome mimetics derived from natural killer cells. Artif Cells Nanomedicine Biotechnol 46:S166–S179. https://doi.org/10.1080/21691401.2018.1489824
6. Redzic JS, Kendrick AA, Bahmed K et al (2013) Extracellular vesicles secreted from cancer cell lines stimulate secretion of MMP-9, IL-6, TGF-β1 and EMMPRIN. PLoS One 8:e71225. https://doi.org/10.1371/journal.pone.0071225
7. Gangadaran P, Li XJ, Lee HW et al (2017) A new bioluminescent reporter system to study the biodistribution of systematically injected tumor-derived bioluminescent extracellular vesicles in mice. Oncotarget 109894–109914:8. https://doi.org/10.18632/oncotarget.22493

8. Blackwell RH, Foreman KE, Gupta GN (2017) The role of cancer-derived exosomes in tumorigenicity & epithelial-to-mesenchymal transition Cancers:9, pii: E105. https://doi.org/10.3390/cancers9080105
9. Gangadaran P, Ahn B-C (2017) Molecular imaging: A useful tool for the development of natural killer cell-based immunotherapies. Front Immunol 8:1090. https://doi.org/10.3389/fimmu.2017.01090
10. Lorenz WW, McCann RO, Longiaru M, Cormier MJ (1991) Isolation and expression of a cDNA encoding *Renilla reniformis* luciferase. Proc Natl Acad Sci U S A 88:4438–4442
11. Hori K, Wampler JE, Matthews JC, Cormier MJ (1973) Identification of the product excited states during the chemiluminescent and bioluminescent oxidation of *Renilla* (sea pansy) luciferin and certain of its analogs. Biochemistry 12:4463–4468
12. Matthews JC, Hori K, Cormier MJ (1977) Substrate and substrate analogue binding properties of *Renilla* luciferase. Biochemistry 16:5217–5220
13. Matthews JC, Hori K, Cormier MJ (1977) Purification and properties of *Renilla reniformis* luciferase. Biochemistry 16:85–91

Chapter 15

Bioluminescence Imaging in the Chick Chorioallantoic Membrane Assay

Benedict Jefferies, Zhichao Tong, and Roman Nawroth

Abstract

For the analysis of tumorigenesis and therapeutic intervention, high throughput technologies that allow the detection of tumor size in the context of a living organism are of need. Here we describe the use of a chorioallantoic membrane model in the developing chick embryo on which growth of a tumor xenograft can be monitored over time, enabling bioluminescence technology.

Key words Chorioallantois membrane model, Tumor xenograft, Bioluminescence, Luciferase, In vivo imaging

1 Introduction

The chorioallantoic membrane (CAM) assay is an established model for the short-term grafting and subsequent study of tumorigenesis, therapy response and the vascularization of tissue grafted on the CAM of a wide variety of specimens in an in vivo environment [1–7]. This, for once, includes tumor cells of varying origin, such as lymphoma [8], sarcoma [9], prostate cancer [10], or melanoma [11], also fibroblasts [4] and stem cells [12]. Also, organ grafts, such as murine limb grafts, liver allografts, human bone, and connective tissue samples have been described to be successfully cultivated on the CAM surface, to subsequently facilitate the CAM vascular structure and to initiate graft regeneration [13–15]. The use of different scaffold materials on the CAM has also been reported and allows for screening of their biocompatibility or role in the vascularization of the graft, suggesting a great potential for this model in the field of tissue engineering [16–18]. The CAM model may be performed in ovo or as a shell-less ex ovo assay, the latter requiring the embryo and any other content of the egg to be transferred to an external cultivation vehicle (e.g., a petri dish). The ex ovo cultivation may present benefits over the in ovo cultivation,

Steven Ripp (ed.), *Bioluminescent Imaging: Methods and Protocols*, Methods in Molecular Biology, vol. 2081,
https://doi.org/10.1007/978-1-4939-9940-8_15, © Springer Science+Business Media, LLC, part of Springer Nature 2020

such as better accessibility to the CAM and the embryo, improved possibilities of angiogenesis studies and grafting of limb buds [14]. Still, in ovo cultivation remains the most widely used protocol for CAM assay, as it is less challenging to handle and sufficient for most research purposes [14].

Many aspects set this model ahead of other in vivo models (e.g., murine or zebra fish models). For once, the chick embryo, at the early stage of development until day 17, is devoid of a mature immune system and hence does not interfere in the grafting process by ways of an immunological response toward the graft [15, 19, 20]. Furthermore, the model enables easy daily visual inspection of the experimental site.

The CAM model represents a cost-effective, quick, and reproducible intermediate between 2D cultures and experiments in rodents and raises few ethical or legal concerns. These aspects, in addition to the potentially large number of specimens that may be studied in a short interval, make the CAM assay an efficient model system.

However, despite advances in utilizing the CAM assay in cancer research, noninvasive approaches to quantifiably measure tumor graft growth dynamics are sparse. Microcomputed tomography has successfully been applied to the CAM assay [21]. While it presents a safe method to quantify growth with seemingly no adverse effects on the development of the embryo or the CAM in respect to radiation, access to a device may be limited and therefore restrict research possibilities [22].

We here present an in ovo approach to the CAM assay, in which bioluminescence imaging technology is applied for further evaluation of tumor growth dynamics on the CAM. This was achieved by grafting cells expressing firefly luciferase, inoculation of grafted cells at set intervals with d-Luciferin and subsequent bioluminescence imaging. Daily analysis of any increase or decrease in growth of the grafted tumors can be demonstrated over the course of 7 days [7]. Quantification was carried out using ImageJ and correlated with other methods for tumor quantification, including weighing of the tumors [6, 7]. Presence of tumor cells in the CAM stroma was verified posthumously via histological sectioning and staining [7].

BLI of CAM grafts may be combined with other established protocols in the chick embryo, such as digital microscopy, fluorescence imaging, immunohistochemical staining, micro-CT or -MRI without hindrance and thus provide a wide range of in vivo application.

2 Materials

2.1 Cell Lines and Eggs

1. Cancer derived cell line that constitutively expresses a firefly luciferase. Here, we used RT112-luc cells that are derived from bladder cancer specimen.
2. Fertilized white Leghorn chicken eggs were purchased from a commercial chicken farm.

2.2 Incubation of Eggs

1. 37 °C incubator.

2.3 Xenograft Transplantation

1. Micro drill with 1–2 mm drill bit.
2. Lens cleansing paper (autoclaved) and absorbent paper.
3. Adhesive tape.
4. Scissors, extra delicate, curved, 10 cm.
5. Silican O-ring with around 4 mm inner diameter.
6. Matrigel (Corning Matrigel Matrix, catalog number 354234).

2.4 Bioluminescence Detection

1. D-Luciferin, potassium salt (SIGMA, 30 mg/ml dissolved in PBS and stored at −20 °C).
2. In vivo imaging system for detection of bioluminescence (for example the IVIS spectrum in vivo imaging system from PerkinElmer).
3. Software for analysis of the bioluminescence data. We use Simple PCI (Imaging Systems, Compix Inc. Cranberry Township, PA, USA).

3 Methods

3.1 Establishment of the Chicken Chorioallantoic Membrane System

All procedures were performed with sterilized tools and materials to avoid contamination. Moving of the eggs should be done carefully to minimize mortality of the embryos.

1. Eggs were incubated at 37 °C directly after shipment and placed with the smaller convexity pointing downward from ED1 (Embryo development day) to ED4 (*see* **Note 2**).
2. At ED4, eggs were gently reversed to the smaller convexity pointing upward and placed back to the incubator. One hour later a hole was drilled through the smaller convexity pointing of the shell without breaking the inner membrane to prevent the eggshell debris to contaminate the inner part of the egg. To further avoid the interior of the egg from shell fragments, adhesive tapes were applied. Eggs were opened carefully with scissors through the hole and also the inner membrane to

create a round window with approximate 5 mm diameter. Then the windows were sealed with surgical tape and the eggs were incubated for another 3 days without disturbance.

3. At ED7 the windows from ED4 were enlarged to 10–15 mm diameter closed again with adhesive tape and incubated until ED8.

3.2 Establishing of a Tumor Xenograft

1. On ED8 the developing CAM will be visible in good developing eggs. For better engraftment, a piece of lens paper can be used to irritate the CAM and create a lacerated wound with micro-hemorrhage (Fig. 1a). Be careful where to place the lens paper because direct irritation of the main blood vessel could lead to exsanguination and thus should be averted (Fig. 1b). In case of too much extra liquid in the egg one can use an absorbent paper to remove excess liquid which would complicate irritation of the CAM.
2. A silicon ring with 4 mm inner diameter is then placed on the site of the generated hemorrhage. Wait approximately 1 h to let the ring settle smoothly on the CAM.
3. 2×10^6 RT112-luc cells resuspended in 10 μl medium were mixed with an additional 10 μl Matrigel and seeded in the inner space of the ring. Air bubbles should be avoided when mixing, since this is the main reason for uneven distribution of tumor cells. The number of cells should be optimized dependent on

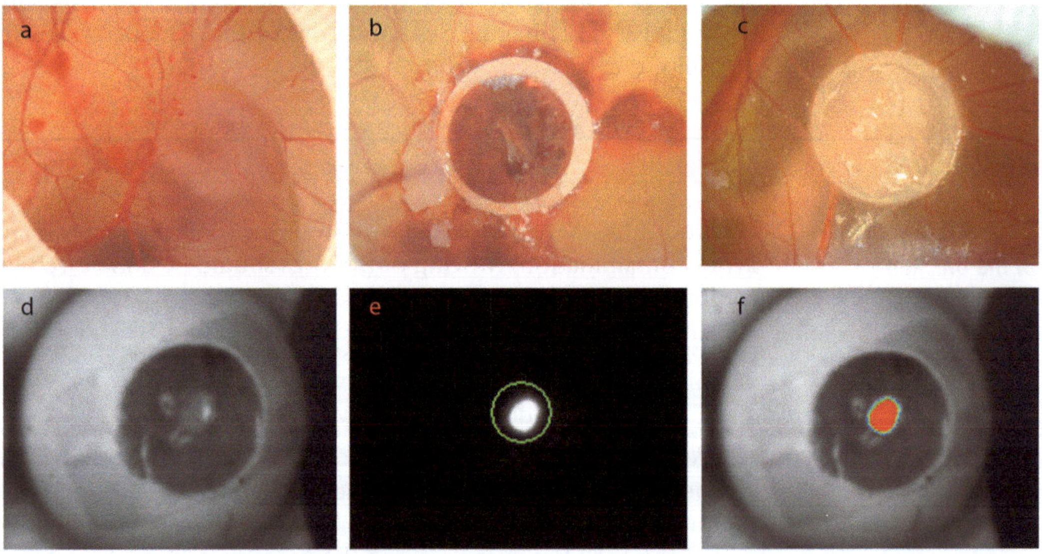

Fig. 1 Establishment and size evaluation of a tumor xenograft in the CAM model using bioluminescence. (**a**) microhemorrhage on CAM, (**b**) exsanguination caused by improper irritation, (**c**) well developed xenograft on CAM, (**d**) Brightfield image of tumor xenograft, (**e**) bioluminescence image of tumor xenograft, (**f**) merged image with Simple PCI

the cell line and the purpose of the assay (*see* **Note 1**). After implantation the window in the egg shell is covered again with adhesive tape and the eggs are placed in the incubator.

4. Tumor formation is examined on ED10 and xenografts are randomized to the vehicle and treated groups after removing dead embryos and unevenly grown xenografts (Fig. 1c).
5. For treatment of tumors, small molecule inhibitors can be applied topically on the tumor in a total volume of 10 μl. The drug concentration can be calculated according to the total blood volume of the developing embryo (1 ml at ED10).

3.3 Bioluminescence Imaging (BLI)

1. On ED15 or 16, 30 μl of the D-luciferin solution is administered directly onto the xenografts. After 15 min incubation at room temperature, brightfield and bioluminescence images (Fig. 1d, e) are acquired. The exposure time largely depends on the sensitivity of the camera used. In our case, we used an IVIS Spectrum system (PerkinElmer) which needs less than a second to acquire an image. However, we could also use a Chemidoc imaging system (Bio-Rad) that did need 15 min exposure which delivered perfectly.
2. Further mergence and quantification (Fig. 1f) are conducted with Simple PCI (Imaging Systems, Compix Inc. Cranberry Township, PA, USA).

3.4 Optional Tumor Harvesting

1. After the imaging process, tumors are cut from the CAM and trimmed under a stereomicroscope to ensure that the attached CAM system was removed. Tumor weight is measured in preweighed microcentrifuge tubes containing 200 μl PBS. For further analysis, the tumor tissue can be freshly frozen or fixed in formalin.

4 Notes

1. The size of forming xenografts depends largely on the cell line used and often is independent of the tumor stage of the origin of the cell line. Before deciding for a particular cell system, a thorough analysis of suitable cell lines should be performed with respect to growth, invasiveness or also vasculogenesis. As for quantification of tumor size by detection of bioluminescence, cells have to express constitutively a firefly luciferase. In our laboratory we are using the bladder cancer derived cell line RT112 transduced with a lentiviral vector encoding for firefly luciferase (RT112-luc). Cells were maintained in RPMI-1640 medium in 5% CO_2 supplemented with 10% FBS, 1% Penicillin/Streptomycin, and 1% nonessential amino acids.

2. As for the eggs, it should be ensured that there is a certified day of fertilization that has to be the same for all eggs and that the eggs have been stored at 4–8 °C. Although up to 2 weeks of storage at this temperature is feasible, survival rates are much higher and also development of the CAM throughout the eggs is more even when the eggs are used as early as possible after fertilization.

References

1. Kain KH, Miller JW, Jones-Paris CR, Thomason RT, Lewis JD, Bader DM, Barnett JV, Zijlstra A (2014) The chick embryo as an expanding experimental model for cancer and cardiovascular research. Dev Dyn 243 (2):216–228. https://doi.org/10.1002/dvdy.24093
2. Ribatti D (2014) The chick embryo chorioallantoic membrane as a model for tumor biology. Exp Cell Res 328(2):314–324. https://doi.org/10.1016/j.yexcr.2014.06.010
3. Tufan AC, Satiroglu-Tufan NL (2005) The chick embryo chorioallantoic membrane as a model system for the study of tumor angiogenesis, invasion and development of anti-angiogenic agents. Curr Cancer Drug Targets 5(4):249–266
4. Mangir N, Raza A, Haycock JW, Chapple C, Macneil S (2018) An improved in vivo methodology to visualise tumour induced changes in vasculature using the chick chorionic allantoic membrane assay. In vivo 32(3):461–472. https://doi.org/10.21873/invivo.11262
5. Kunzi-Rapp K, Genze F, Kufer R, Reich E, Hautmann RE, Gschwend JE (2001) Chorioallantoic membrane assay: vascularized 3-dimensional cell culture system for human prostate cancer cells as an animal substitute model. J Urol 166(4):1502–1507
6. Skowron MA, Sathe A, Romano A, Hoffmann MJ, Schulz WA, van Koeveringe GA, Albers P, Nawroth R, Niegisch G (2017) Applying the chicken embryo chorioallantoic membrane assay to study treatment approaches in urothelial carcinoma. Urol Oncol 35(9):544 e511–544 e523. https://doi.org/10.1016/j.urolonc.2017.05.003
7. Jefferies B, Lenze F, Sathe A, Truong N, Anton M, von Eisenhart-Rothe R, Nawroth R, Mayer-Kuckuk P (2017) Non-invasive imaging of engineered human tumors in the living chicken embryo. Sci Rep 7(1):4991. https://doi.org/10.1038/s41598-017-04572-1
8. Klingenberg M, Becker J, Eberth S, Kube D, Wilting J (2014) The chick chorioallantoic membrane as an in vivo xenograft model for Burkitt lymphoma. BMC Cancer 14:339. https://doi.org/10.1186/1471-2407-14-339
9. Sys GM, Lapeire L, Stevens N, Favoreel H, Forsyth R, Bracke M, De Wever O (2013) The in ovo CAM-assay as a xenograft model for sarcoma. J Vis Exp 77:e50522. https://doi.org/10.3791/50522
10. Fang J, Ding M, Yang L, Liu LZ, Jiang BH (2007) PI3K/PTEN/AKT signaling regulates prostate tumor angiogenesis. Cell Signal 19 (12):2487–2497. https://doi.org/10.1016/j.cellsig.2007.07.025
11. Busch C, Krochmann J, Drews U (2013) The chick embryo as an experimental system for melanoma cell invasion. PLoS One 8(1): e53970. https://doi.org/10.1371/journal.pone.0053970
12. Handel M, Hammer TR, Nooeaid P, Boccaccini AR, Hoefer D (2013) 45S5-bioglass((R))-based 3D-scaffolds seeded with human adipose tissue-derived stem cells induce in vivo vascularization in the CAM angiogenesis assay. Tissue Eng Part A 19(23–24):2703–2712. https://doi.org/10.1089/ten.TEA.2012.0707
13. Katoh M, Nakada K, Miyazaki JI (2001) Liver regeneration on chicken chorioallantoic membrane. Cells Tissues Organs 169(2):125–133. https://doi.org/10.1159/000047870
14. Dohle DS, Pasa SD, Gustmann S, Laub M, Wissler JH, Jennissen HP, Dunker N (2009) Chick ex ovo culture and ex ovo CAM assay: how it really works. J Vis Exp 33(33). https://doi.org/10.3791/1620
15. Martinez-Madrid B, Donnez J, Van Eyck AS, Veiga-Lopez A, Dolmans MM, Van Langendonckt A (2009) Chick embryo chorioallantoic membrane (CAM) model: a useful tool to study short-term transplantation of cryopreserved human ovarian tissue. Fertil Steril 91

(1):285–292. https://doi.org/10.1016/j.fertnstert.2007.11.026

16. Baiguera S, Macchiarini P, Ribatti D (2012) Chorioallantoic membrane for in vivo investigation of tissue-engineered construct biocompatibility. J Biomed Mater Res B Appl Biomater 100(5):1425–1434. https://doi.org/10.1002/jbm.b.32653
17. Ribatti D, Vacca A, Roncali L, Dammacco F (2000) The chick embryo chorioallantoic membrane as a model for in vivo research on anti-angiogenesis. Curr Pharm Biotechnol 1 (1):73–82
18. Conconi MT, Nico B, Rebuffat P, Crivellato E, Parnigotto PP, Nussdorfer GG, Ribatti D (2005) Angiogenic response induced by acellular femoral matrix in vivo. J Anat 207 (1):79–83. https://doi.org/10.1111/j.1469-7580.2005.00427.x
19. Weber WT, Mausner R (1977) Migration patterns of avian embryonic bone marrow cells and their differentiation to functional T and B cells. Adv Exp Med Biol 88:47–59
20. Jankovic BD, Isakovic K, Markovic BM, RAjcević M (1977) Immunological capacity of the chicken embryo. II. Humoral immune responses in embryos and young chickens bursectomized and sham-bursectomized at 52--64 h of incubation. Immunology 32 (5):689–699
21. Kim JS, Min J, Recknagel AK, Riccio M, Butcher JT (2011) Quantitative three-dimensional analysis of embryonic chick morphogenesis via microcomputed tomography. Anat Rec (Hoboken) 294(1):1–10. https://doi.org/10.1002/ar.21336
22. Henning AL, Jiang MX, Yalcin HC, Butcher JT (2011) Quantitative three-dimensional imaging of live avian embryonic morphogenesis via micro-computed tomography. Dev Dyn 240 (8):1949–1957. https://doi.org/10.1002/dvdy.22694

Chapter 16

Application of a cybLuc Aminoluciferin for Deep Tissue Bioluminescence Imaging in Rodent Models

Xiang Li and Minyong Li

Abstract

Bioluminescent imaging (BLI) technology has been extensively applied due to various advantages such as noninvasiveness, high sensitivity and selectivity, excellent biocompatibility and real-time visualization and monitoring. The firefly luciferase (Fluc)/luciferin system, one of the principal bioluminescent systems, has been developed as a sensor for imaging biological processes. However, a limited number of Fluc substrates hamper the further application of firefly luciferase/luciferin systems for biomedical purposes. Here we describe an approach to synthesize a series of novel luciferin substrates (cyaLucs) that produced elevated bioluminescent signals in vitro. Furthermore, we demonstrate the high efficiency of *N*-cyclobutylamino-luciferin (cybLuc) with high light emission and long duration in deep tissue imaging by diagnosis of cerebral tumors in vivo in a rodent model.

Key words cybLuc aminoluciferin derivative, Bioluminescence imaging, Deep tissue imaging, Rodent model

1 Introduction

Bioluminescent imaging (BLI) is a consistently sensitive, specific, simple, and noninvasive imaging technique that has been comprehensively applied for monitoring a myriad of life processes, including cell proliferation and migration, tumor growth, enzyme activities, protein–protein interactions, and other uses [1]. Among the various bioluminescent systems, the firefly luciferase–luciferin system has been wildly applied as highly bioluminescent sensors [2]. The selective and specific interaction between luciferin-based substrates and associated enzymes limits the development of novel luciferase substrates. The majority of BLI applications are largely dependent on the wild-type substrate D-luciferin (dLuc) or the analog, aminoluciferin (aLuc). Thus, there is increasing demand for novel light-emitting Fluc substrates [3].

Currently, several key unfavorable factors have hindered the in vivo employment of BLI systems. One is that bioluminescent

Steven Ripp (ed.), *Bioluminescent Imaging: Methods and Protocols*, Methods in Molecular Biology, vol. 2081, https://doi.org/10.1007/978-1-4939-9940-8_16, signals below 600 nm attenuate strongly and rapidly because deep tissues are able to absorb and scatter the light [4]. This hampers the extension of BLI technology to deep tissues as well as large animals [5–10]. Another limitation is that the circulation time of luciferin is relatively short in vivo. In addition, moderate permeability of the majority of luciferins is a considerable deficiency for BLI to be applied in vivo. Although several luciferins have already been employed for imaging in vitro and in vivo, none of them are capable of deep tissue imaging.

In order to overcome the disadvantages of luciferins for deep tissue imaging, certain strategies have been proposed. Some modified analogs of dLuc have been developed to obtain robust red wavelength shifting and higher emission of light [5–12]. These substrates initially create intense bioluminescence that subsequently drops to a much lower level. Some studies have prolonged the half-life of luciferins to some extent, but chemical preparations are too complicated to apply [6, 13]. Currently, only a chemical strategy designed to enhance permeability by increasing the lipophilicity and reducing the polarity of the molecule has proven feasible [14–17].

Herein, on the basis of this strategy, we further proposed a series of novel luciferins (*N*-cycloalkylaminoluciferins, cyaLucs) for effective bioluminescence. These potential candidates were prepared by introducing lipophilic *N*-cycloalkyl groups onto aLuc. Cell permeability, bioluminescence sensitivity, and brain imaging were significantly increased. In addition, evaluation in vitro and in vivo revealed superior bioluminescent properties as compared with those of dLuc and aLuc. In sum, we have synthesized cyaLucs with enhanced cell permeability, elevated red wavelength-shifting and overall light emission and prolonged circulatory life. Importantly, *N*-cyclobutylaminoluciferins (cybLucs) are capable of deep tissue imaging and have been successfully utilized for diagnosis of cerebral tumors in vivo.

2 Materials

All reagents and equipment were obtained from commercial sources and were used as received, unless otherwise noted.

2.1 Synthesis of Substrates

1. Milli-Q water: water used for the preparation of aqueous solutions was obtained from a Millipore water purification system (>18.2 MΩ, Milli-Q, Millipore).
2. NMR spectra were obtained in deuterated solvents on Bruker AV-300 spectrometers (*see* **Note 1**).

2.2 Bioluminescence Detection

1. Tris–HCl buffer containing luciferase ATP to solutions of the substrate at various concentrations.
2. ES-2-Fluc cells passed and plated in 96-well black plates with clear bottoms.
3. Balb/c nude mice, 8 weeks of age, were purchased from the Animal Center of the China Academy of Medical Sciences (Beijing, China).
4. Pathogen-free luciferase-expressing transgenic mice (FVB-Tg (CAG-luc, -GFP) L2G85Chco/FathJ17).
5. Hitachi F4500 fluorescence spectrophotometer (Hitachi High Technologies America, Inc., Schaumburg, Illinois, USA) with a blocked excitation path at 37 °C was used to collect the bioluminescence.
6. IVIS Kinetic imaging system (PerkinElmer, Hopkinton, Massachusetts, USA) equipped with a cooled CCD camera was used to obtain bioluminescent images.
7. Hamilton syringe (180 μm needle) using a micropump system with flow rates of 1.500 nL/min (withdrawal) and 500 nL/min (injection) was used for brain injection.

3 Methods

3.1 General Synthetic Route for a Series of N-Cycloalkylaminoluciferins (cyaLucs)

The synthetic route of cyaLucs is generally depicted as in Fig. 1. Detailed synthesis processes are performed according to previously reported methods [18] (*see* **Note 2**).

1. Use acetic acid as a solvent and catalyst to obtain of 6-amino-benzo[d]thiazole-2-carbonitrile (compound 13) in the one-pot reaction.
2. Use a column to separate the reaction products and then get pure monocycloalkyl intermediates **14a-c.**

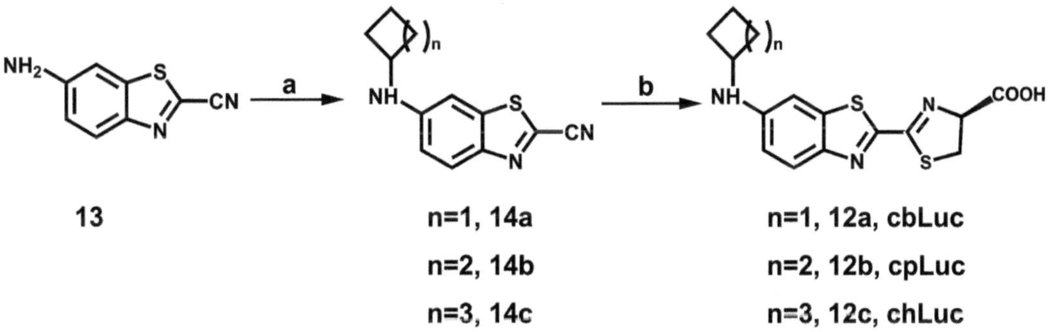

Fig. 1 Synthesis of *N*-cycloalkylaminoluciferins (cyaLucs). (**a**) cyclanone, acetic acid, sodium cyanoborohydride; (**b**) D-cysteine hydrochloride, potassium carbonate (K_2CO_3), $DCM/CH_3OH/H_2O$, N_2, room temperature

3. Under darkness and N_2 atmosphere, cross-coupling reaction D-cysteine hydrochloride and monocycloalkyl intermediates **14a-c** produce the cyaLucs (12a, cy-bLuc; 12b, cypLuc; 12c, cyhLuc).

3.2 Synthesis of 6-(Cyclobutylamino) Benzo[d]Thiazole-2-Carbonitrile (14a)

1. Dissolve 50 mg (0.29 mmol) 5-aminobenzo[d]thiazole-2-carbonitrile (compound 13) in 1 mL acetic acid, then 187.6 mg (2.68 mmol) cyclobutanone and 62 mg (1 mmol) cyanoborohydride is added. Stir the mixture until the solution becomes clear.
2. Dissolve resulting yellow solid with ethyl acetate, then wash the organic phase three times with saturated Na_2CO_3 solution and for twice with brine. Use sodium sulfate to dry over and concentrate.
3. Purify the crude material on a silica column under reduced pressure to afford 40 mg of pure product (**14a**) as a yellow powder (62.5%).

3.3 Synthesis of (S)-2-(6-(Cyclobutylamino) Benzo[d]Thiazol-2-Yl)-4,5-Dihydrothiazole-4-Carboxylic Acid (12a, cybLuc)

1. Dissolve compound 14 (30 mg, 0.13 mmol) in 2 mL N_2-sparged mixture solution of dichloromethane (DCM): CH_3OH (1:1).
2. Add D-cysteine hydrochloride (46 mg, 0.26 mmol) and potassium carbonate (35.9 mg, 0.26 mmol) in 2.5 mL N_2-sparged DI-H_2O–MeOH (1.5:1) under an N_2 atmosphere.
3. Keep the reaction away from light for 10 min while under stirring.
4. Remove the DCM and MeOH under low pressure, then regulate the pH value of the aqueous solution by adding HCl (1 M) solution to pH 6 until a precipitate forms.
5. Subsequently, filter the precipitate off and dry under reduced pressure to afford 24 mg of pure product 12a (cybLuc) as a yellow powder (56%). Purity: >95%.

3.4 Synthesis of 6-(Cyclohexylamino) Benzo[d]Thiazole-2-Carbonitrile (14c)

1. Dissolve 5-aminobenzo[d]thiazole-2-carbonitrile (compound 13, 50 mg, 0.29 mmol) in 1 mL acetic acid.
2. Add cyclobutanone (50 mg, 0.5 mmol) and sodium cyanoborohydride (62 mg, 1 mmol).
3. Stir the mixture at 0 °C until the solution becomes clear.
4. Dissolve the yellow solid in ethyl acetate, then wash the organic phase three times with saturated Na_2CO_3 solution, wash twice with brine, and dry over sodium sulfate to concentrate.
5. Purify the crude material on a silica column dried under reduced pressure to afford 50 mg of pure product (compound **14c**) as a yellow powder (68%).

3.5 Synthesis of (S)-2-(6-(Cyclohexylamino) Benzo[d]Thiazol-2-Yl)-4,5-Dihydrothiazole-4-Carboxylic Acid (12c, cyhLuc)

1. Dissolve compound 14c (30 mg, 0.12 mmol) in a 2 mL N_2-sparged mixture solution of DCM:CH_3OH (1:1).
2. Add D-cysteine hydrochloride (42 mg, 0.24 mmol) and potassium carbonate (33 mg, 0.24 mmol) dissolved in 2.5 mL N_2-sparged DI-H_2O–MeOH (1.5:1) under an N_2 atmosphere.
3. Stir the reaction for 10 min away from light.
4. Remove the DCM and MeOH under low pressure.
5. Add HCl (1 M) solution to set the pH value of the aqueous solution to 6.
6. Filter the precipitate off and dry under reduced pressure to afford 18 mg of pure product 14c (cyhLuc) as an orange-red powder (56%). Purity: >95%.

3.6 In Vitro Bioluminescence Imaging

1. Add 50 μL Tris–HCl buffer with 20 μg/mL luciferase and 2 mM ATP to the solutions of the substrate at various concentrations (0.01–1 μM) in Tris buffer (50 μL).
2. Set the acquisition time of the IVIS instrument to 0.5 s and detect the bioluminescent signals (Fig. 2a).
3. Add 50 μL substrate (20 μM) to various concentrations of ATP (0.25–10 μM) solutions in Tris–HCl buffer containing 20 μg/mL luciferase (50 μL). Set the IVIS acquisition time to 0.5 s and detect the bioluminescent signals (Fig. 2b).

3.7 Bioluminescence Cell Imaging

1. Pass and plate ES-2-Fluc cells in 96-well black plates (4×10^4 cells per well) or at various concentrations (1250, 2500, 5000, 10,000, 20,000, and 40,000 cells per well) with clear bottoms.

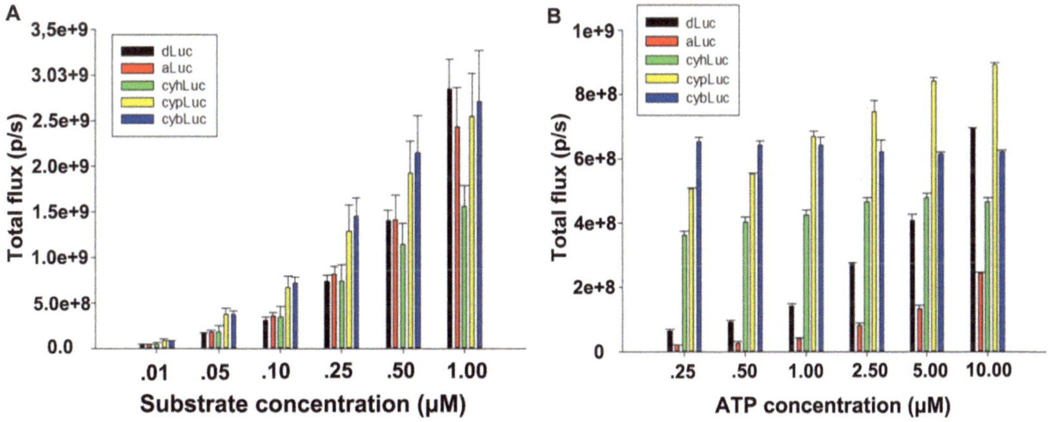

Fig. 2 Dose–response analysis of substrates (**a**) and ATP (**b**). (**a**) (0.01–1 μM) substrates incubated with 10 μg/mL native luciferase in 50 mM Tris–HCl buffer containing 10 mM $MgCl_2$, 0.1 mM $ZnCl_2$, and 1 mM ATP (pH 7.4); (**b**) (0.25–10 μM) ATP incubated with 10 μg/mL native luciferase in 50 mM Tris–HCl buffer containing 10 mM $MgCl_2$, 0.1 mM $ZnCl_2$, and 10 μM luciferins (pH 7.4)

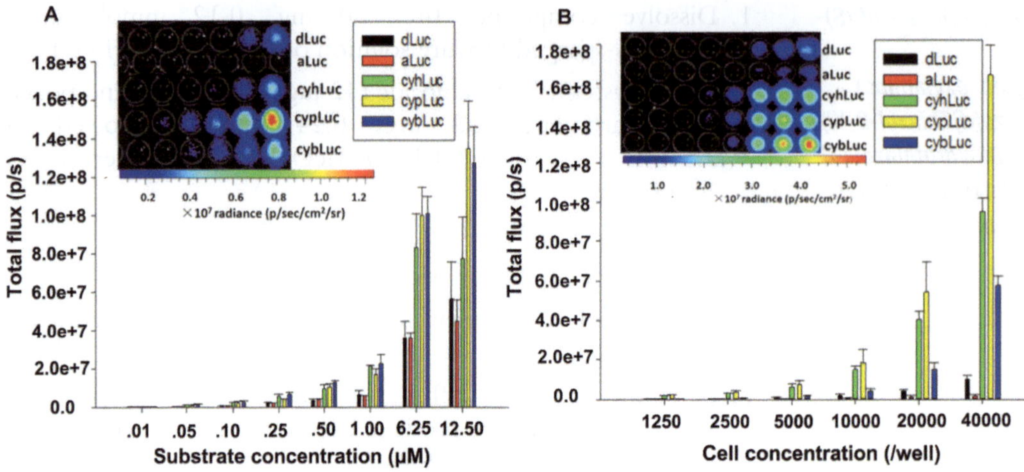

Fig. 3 Dose-response bioluminescence analysis of substrates (**a**) and cells (**b**); (**a**) bioluminescence imaging of (0.01–12.50 μM) substrates incubated with ES-2-Fluc cells (4×10^4 cells per well) and quantification of the bioluminescent imaging signal; (**b**) bioluminescence imaging of substrates incubated with various concentrations of ES-2-Fluc cells ($1.25 \times 10^3 - 4 \times 10^4$ per well) and quantification of the bioluminescent imaging signals. All assays were performed in triplicate and presented as the mean ± SEM

2. Remove the medium when the cells become approximately 95% confluent. Then add various concentrations of substrate in normal saline (NS).
3. Subsequent to the addition of the substrates, measure the bioluminescence with an IVIS acquisition time of 1 or 20 s (Fig. 3a) (*see* **Note 3**).
4. Pass ES-2-Fluc cells and plate at various concentrations (1250, 2500, 5000, 10,000, 20,000, and 40,000 cells per well) in black 96-well plates with clear bottoms.
5. When the cells become approximately 95% confluent, aspirate off the medium and add 50 μL of substrate (20 μM) in NS to each well.
6. Measure the bioluminescence immediately after the addition of the substrates with an IVIS acquisition time of 1 s (Fig. 3b) (*see* **Note 3**).

3.8 In Vivo Bioluminescence Imaging

1. Obtain tumor xenografted mice by subcutaneously implanting ES-2-Fluc cells (1×10^7) under the right forelimb armpit of each 6–8-week-old nude mouse (*see* **Note 4**).
2. Harvest the tumor and cut into 10 mg pieces, and then implant 10-mg tumor pieces subcutaneously into the right armpit region (*see* **Note 5**).
3. Anesthetize the mice baring ES-2-Fluc subcutaneous tumors with isoflurane and inject them with 100 μL of various

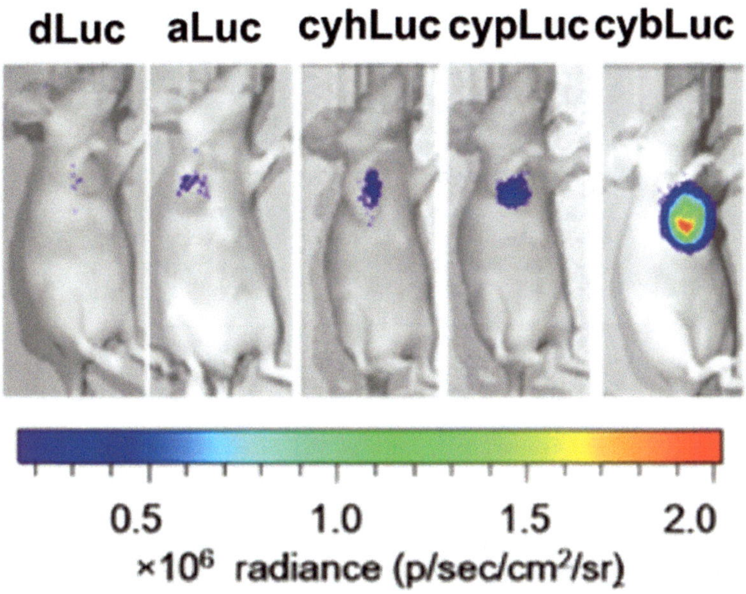

Fig. 4 Representative bioluminescence images of a nude mouse implanted with ES-2-Fluc xenografts after intraperitoneal injections of various luciferins (10 μM, 0.1 mL)

concentrations of substrate (10 μM, 100 μM, 1 mM, 4 mM, and 10 mM).

4. After 10 min, acquire bioluminescent images at various acquisition times (60, 30, 20, 10 and 1 s), and again at 6 ± 1 min until the intensity stabilizes. When the signal reaches a plateau, measure the intensity using a cy5.5 filter (Fig. 4).
5. Inject the dLuc, aLuc, and cybLuc substrates (100 μL of 1 mM solutions in NS, intravenously (i.v.) into luciferase-expressing FVB transgenic mice (*see* **Note 6**).
6. Acquire the bioluminescent images with an acquisition time of 1 s and again at 5 ± 1 min until the intensity stabilizes. The heads and backs of the mice are the regions of interests (ROIs), ROI 1 and ROI 2, respectively. Use the ratio of ROI 1 and ROI 2 as an index for crossing of the blood–brain barrier.
7. Maintain the ES-2-Fluc cell suspensions on ice during surgery.
8. Subsequently, inject cell suspensions into the brain with a Hamilton syringe (180 μm needle) using a micropump system with flow rates of 1.500 nL/min (withdrawal) and 500 nL/min (injection) after leaving the needle in place for 2 min.
9. After 12 days, anesthetize mice with ES-2-Fluc subcutaneous tumors with isoflurane and inject with dLuc, aLuc, or cybLuc (intraperitoneally (i.p.), 1 mM, 200 μL).

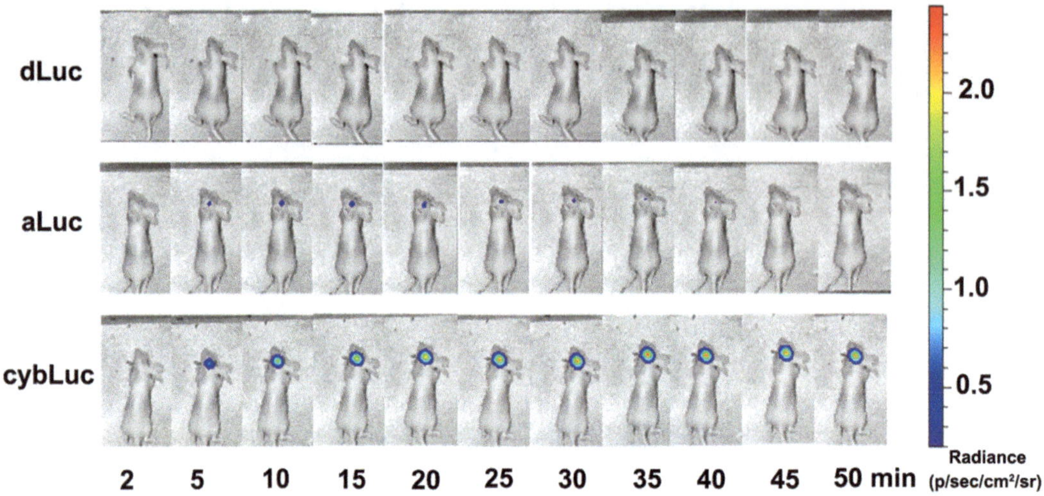

Fig. 5 Bioluminescence imaging of substrates with ES-2-Fluc in nude mouse brain. Representative bioluminescence images after intraperitoneal injection of dLuc, aLuc, and cybLuc (1 mM, 0.2 mL) over time

10. Acquire bioluminescent images with a 10-s acquisition time and again at 5 min intervals until the intensity stabilizes (Fig. 5).

4 Notes

1. All chemical shifts are reported in the standard δ notation of parts per million using the peaks of residual proton and carbon signals of the solvent as internal references. NMR peaks are referred to as singlet (s), doublet (d), doublet of doublets (dd), triplet (t), or broad singlet (br). Coupling constants (J) are reported in Hertz. Mass spectra are recorded in ESI^+ mode (70 eV).
2. The details for the preparation of all substrates and their NMR and HR-MS spectra can be found in the Supplementary Materials of the Wu et al. (2017) paper [18].
3. Use a cooled CCD camera such as the IVIS instrument to collect photon emissions.
4. Ensure that all animal studies are approved by your institutions Ethics Committee. Work performed by the authors has been approved by the Ethics Committee and IACUC of Qilu Health Science Center, Shandong University, and were conducted in compliance with European guidelines for the care and use of laboratory animals.
5. The tumor xenografts were allowed to grow for 2 weeks before imaging.

6. The mice used were littermates (8 weeks of age, males) and were housed singly or in groups and maintained on a 12:12 h light–dark cycle at 22 °C with free access to food and water.

Acknowledgments

This work was supported by grants from the National Natural Science Foundation of China (No. 81673393), the Taishan Scholar Program at Shandong Province, the Qilu/Tang Scholar Program at Shandong University, the Program for Changjiang Scholars and Innovative Research Team in University (No. IRT_17R68), the Key Research and Development Project of Shandong Province (No. 2017CXGC1401), and the Fundamental Research Funds of Shandong University (No. 2017GN0030) as well as China Post-doctoral Science Foundation (No. 2018 M640640).

References

1. Paley MA, Prescher JA (2014) Bioluminescence: a versatile technique for imaging cellular and molecular features. MedChemComm 5 (3):255–267
2. Wu W, Li J, Chen L, Ma Z, Zhang W, Liu Z, Cheng Y, Du L, Li M (2014) Bioluminescent probe for hydrogen peroxide imaging in vitro and in vivo. Anal Chem 86(19):9800–9806
3. Li J, Chen L, Du L, Li M (2013) Cage the firefly luciferin!–a strategy for developing bioluminescent probes. Chem Soc Rev 42 (2):662–676
4. Negrin RS, Contag CH (2006) In vivo imaging using bioluminescence: a tool for probing graft-versus-host disease. Nat Rev Immunol 6 (6):484
5. White EH, Wörther H, Seliger HH, McElroy WD (1966) Amino analogs of firefly luciferin and biological activity thereof. J Am Chem Soc 88(9):2015–2019
6. Shinde R, Perkins J, Contag CH (2006) Luciferin derivatives for enhanced in vitro and in vivo bioluminescence assays. Biochemistry 45(37):11103–11112
7. Woodroofe CC, Shultz JW, Wood MG, Osterman J, Cali JJ, Daily WJ, Meisenheimer PL, Klaubert DH (2008) N-alkylated 6-′-aminoluciferins are bioluminescent substrates for ultra-Glo and QuantiLum luciferase: new potential scaffolds for bioluminescent assays. Biochemistry 47(39).10383–10393
8. Reddy GR, Thompson WC, Miller SC (2010) Robust light emission from cyclic alkylaminoluciferin substrates for firefly luciferase. J Am Chem Soc 132(39):13586–13587
9. Conley NR, Dragulescu-Andrasi A, Rao J, Moerner W (2012) A selenium analogue of firefly D-luciferin with red-shifted bioluminescence emission. Angew Chem 124 (14):3406–3409
10. McCutcheon DC, Paley MA, Steinhardt RC, Prescher JA (2012) Expedient synthesis of electronically modified luciferins for bioluminescence imaging. J Am Chem Soc 134 (18):7604–7607
11. Mofford DM, Reddy GR, Miller SC (2014) Aminoluciferins extend firefly luciferase bioluminescence into the near-infrared and can be preferred substrates over D-luciferin. J Am Chem Soc 136(38):13277–13282
12. Pirrung MC, Biswas G, De Howitt N, Liao J (2014) Synthesis and bioluminescence of difluoroluciferin. Bioorg Med Chem Lett 24 (20):4881–4883
13. Chandran SS, Williams SA, Denmeade SR (2009) Extended-release PEG–luciferin allows for long-term imaging of firefly luciferase activity in vivo. Luminescence 24(1):35–38
14. Evans MS, Chaurette JP, Adams ST Jr, Reddy GR, Paley MA, Aronin N, Prescher JA, Miller SC (2014) A synthetic luciferin improves bioluminescence imaging in live mice. Nat Methods 11(4):393
15. Kuchimaru T, Iwano S, Kiyama M, Mitsumata S, Kadonosono T, Niwa H, Maki S, Kizaka-Kondoh S (2016) A luciferin

analogue generating near-infrared bioluminescence achieves highly sensitive deep-tissue imaging. Nat Commun 7:11856

16. Mofford DM, Adams ST Jr, Reddy GKK, Reddy GR, Miller SC (2015) Luciferin amides enable in vivo bioluminescence detection of endogenous fatty acid amide hydrolase activity. J Am Chem Soc 137(27):8684–8687
17. Mofford DM, Miller SC (2015) Luciferins behave like drugs. ACS Chem Neurosci 6 (8):1273–1275
18. Wu W, Su J, Tang C, Bai H, Ma Z, Zhang T, Yuan Z, Li Z, Zhou W, Zhang H (2017) cybLuc: an effective aminoluciferin derivative for deep bioluminescence imaging. Anal Chem 89(9):4808

Part IV

In Planta Imaging Applications

Chapter 17

Application of Single-Cell Bioluminescent Imaging to Monitor Circadian Rhythms of Individual Plant Cells

Tomoaki Muranaka and Tokitaka Oyama

Abstract

The bioluminescent reporter system is a powerful tool for the long-term monitoring of gene expression because of its noninvasive nature. Furthermore, in combination with high-sensitive imaging technology, spatiotemporal analysis on regulation and heterogeneity in gene expression is possible. We developed a single-cell bioluminescent imaging system for plants through a transient gene transfection by particle bombardment. By applying this system to a duckweed species, we succeeded in monitoring circadian rhythms of individual cells in an intact plant for over a week. Here we describe methods for gene transfection by particle bombardment and single-cell bioluminescence monitoring by a high-sensitive camera. This technique provides a platform for characterizing gene expression patterns of individual cells in the same tissue.

Key words Bioluminescence, Circadian rhythm, Duckweed, Gene expression, Single cell

1 Introduction

Time-series data of gene expression provide critical information to the studies on gene regulatory architecture. The bioluminescent reporter system is a major tool for in vivo monitoring of gene expression dynamics due to its noninvasive nature and quantitative performance. The application of a bioluminescent reporter system to monitor circadian rhythms was first reported in 1992; bioluminescence images of transgenic tobacco seedlings carrying a luciferase gene under control of a circadian promoter (*CAB2::LUC*) were captured every 4 h by using an intensified camera [1]. Thereafter the in vivo bioluminescence systems have been efficiently used in research on the circadian clock of various organisms such as cyanobacteria, Arabidopsis, Drosophila, mammals, and so on [2–5].

The circadian clock is an endogenous timing system based on self-sustained oscillations with a period of approximately 24 h. The circadian oscillators in eukaryotes consist of complex gene regulatory networks including feedback loops [6]. These gene circuits

Steven Ripp (ed.), *Bioluminescent Imaging: Methods and Protocols*, Methods in Molecular Biology, vol. 2081,
https://doi.org/10.1007/978-1-4939-9940-8_17,

are implemented in individual cells. Thus, the circadian system of a multicellular organism is composed of numerous cellular clocks. Bioluminescent imaging of cultured tissues and cells at a single-cell level has been used in many studies of mammalian circadian systems [7–10]. As an achievement, it has been clearly shown that neurons in suprachiasmatic nucleus (SCN) are coupled with each other to generate a spatiotemporal orchestration [7, 10]. In the studies of plant circadian systems, the bioluminescent reporter system has been frequently used for the phenotyping of Arabidopsis seedlings [11, 12]. In the recent decade, spatiotemporal analyses of bioluminescence rhythms in seedlings, leaves, roots of transgenic plants were performed, and inter- and intra-organ desynchronization was revealed [13, 14]. These desynchronization phenomena implied the weakness of coupling between cellular clocks in the plant circadian system. However, there was little information about cellular circadian behavior because it was difficult to monitor bioluminescence rhythms of transgenic plants at a single-cell level. In transgenic plants expressing a luciferase reporter, bioluminescence of individual cells is indistinguishable in intact tissues due to the overlap of their luminescence signals.

To overcome this problem, we dispersedly transfected cells near the surface of a plant by using particle bombardment. As the plant material for the single-cell bioluminescent imaging, we used a duckweed species, *Lemna gibba* [15]. Duckweeds are a group of monocotyledonous plants that have been used for various physiological studies [16]. Recently, whole genome information of several duckweed species was available [17]. Duckweeds are suitable for high-resolution imaging of an entire plant because of the tiny and flat bodies. In addition, as floating on the water, they are immobile in the vertical direction, making it unnecessary to adjust the focus during a long-term monitoring. We constructed an automated imaging system that was composed of a high sensitive EM-CCD camera and a PC-controlled rotary sample table with an LED illumination system. Using this system, we succeeded in monitoring circadian rhythms of individual cells in duckweeds for over a week [18]. Here, we describe methods of efficient gene transfection by particle bombardment and long-term single-cell bioluminescence monitoring. In principle, these methods can be applied to any plant species. This monitoring system is also useful in functional analysis of circadian clock-related genes by cointroduction of effectors for overexpression, RNA interference, and CRISPR/Cas9 systems [19].

2 Materials

2.1 Plants and Culture Medium

1. *Lemna gibba* plants (*see* **Note 1**) maintained in NF medium with 1% sucrose under constant white light conditions (~50 $\mu Em^{-2}\ s^{-1}$) in a temperature-controlled room or a growth chamber (25° ± 1 °C) (*see* **Note 2**).
2. NF medium: 2.7 mM $CaCl_2$, 1.2 mM $MgSO_4$, 1 mM KH_2PO_4, 5 mM KNO_3, 18 μM $MnCl_2$, 46 μM H_3BO_3, 0.77 μM $ZnSO_4$, 0.32 μM $CuSO_4$, 0.49 μM MoO_3, 20 μM $FeSO_4$, 50 μM Na_2EDTA. Adjust pH to 5.0 with KOH.

2.2 Particle Bombardment System

1. Particle delivery system (PDS-1000/He, Bio-Rad).
2. Vacuum pump (SVR-16F, Hitachi).
3. Helium gas cylinder (>650 psi).
4. Rupture disks (450 psi, Bio-Rad).
5. Macrocarriers, macrocarrier holders, macrocarrier insertion tool (Bio-Rad).
6. Stopping screens, (Bio-Rad).
7. Gold particles (1.0 μm diameter, Bio-Rad). Wash gold particles twice with 70% EtOH then suspend them in 50% glycerol (60 mg/ml). Store at −20 °C.
8. 2.5 M $CaCl_2$. Store at −20 °C.
9. 0.1 M spermidine. Store at −20 °C.
10. Reporter construct [*pUC-AtCCA1:LUC+* (*AtCCA1::LUC*)] (*see* **Note 3**). 1 μg/μl. Store at −20 °C.
11. Filter sterilized 0.1 M D-luciferin, potassium salt. Store at −20 °C.
12. Siliconized microcentrifuge tube.
13. Vortex mixer.
14. Microtube centrifuge.
15. Ultrasonic washing machine (AS ONE Corporation).
16. 60-mm plastic dishes (Iwaki).

2.3 Imaging System

Figure 1 shows an overview of the imaging system.

1. EM-CCD camera (ImagEM C9100-13, Hamamatsu Photonics).
2. Circulating water cooler (NCB-1200, EYELA).
3. Windows desktop PC. Install a CameraLink interface board and an RS232C serial interface board.
4. Imaging software (HoKaWo, Hamamatsu Photonics).

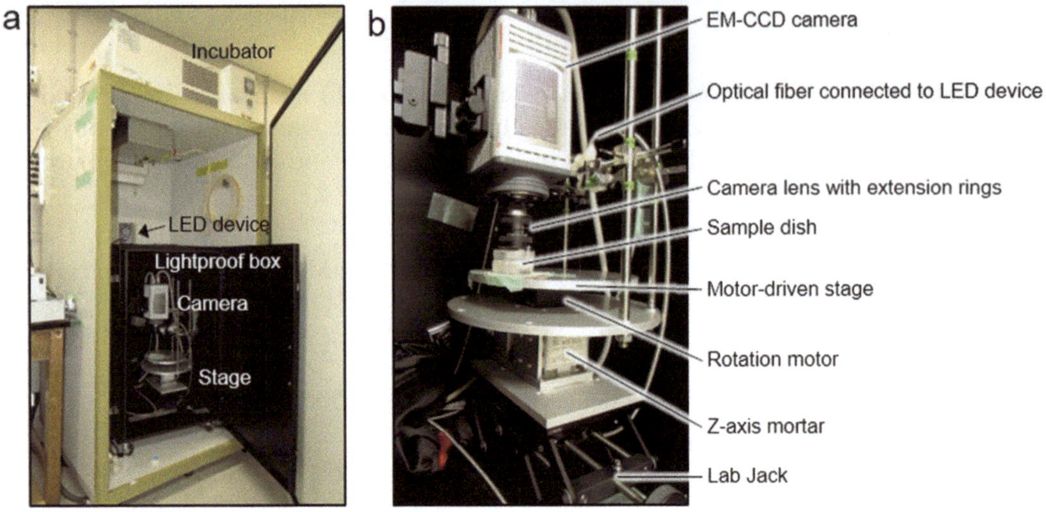

Fig. 1 Overview (**a**) and close-up (**b**) of the imaging system

5. C-mount mega-pixel lens (Xenon 0.95/25-0037, Schneider Optics).
6. Extension and spacer rings for lens (10 mm, 5 mm, 1 mm, and 0.5 mm, Asahi Precision).
7. Short-pass filter to cut delayed autofluorescence from chloroplasts (SV630, Asahi Spectra).
8. LED device for plant illumination (RFB2-20SW, CCS Inc.). Connect to the HoKaWo-installed PC via an RS232C cable.
9. Optical fiber cable (FCB-W, CCS Inc.) guiding illumination from the LED device.
10. Motor-driven stage: a handmade stage composed of aluminum discs, *Z*-axis motor (ALV-902-HP, Chuo Precision Industrial Co.), rotation motor (ARS-6036-GM, Chuo Precision Industrial Co.) (*see* **Note 4**).
11. Motor controller (QT-ADM3, Chuo Precision Industrial Co.) Connect to the motors via ACB-STB3/STM3 cables (Chuo Precision Industrial Co.) and the HoKaWo-installed PC via an RS232C cable.
12. Precision Lab Jack (LJA-16223, SIGMAKOKI).
13. Camera stand (SL700, SFC).
14. Custom-made lightproof box (size: Depth 540 mm, Width 565 mm, Height 870 mm).
15. Incubator (KCLP-1000I, NK-system).
16. 35-mm plastic dishes (Iwaki).
17. Silicone rubber sheet (1 mm thick, 10 × 10 mm).
18. Insect pins (Stainless steel, 0.1 mm diameter, Entomoravia).

3 Methods

3.1 Coating Gold Particles with DNA

1. Suspend gold particles in 50% glycerol by vigorous pipetting (*see* **Note 5**).
2. Transfer an 8-μl aliquot of gold particle suspension to a 1.5-ml siliconized microcentrifuge tube.
3. Set the tube to a vortex mixer.
4. While vortexing, add 2 μl of a reporter plasmid (1 μg/μl), 3.3 μl of 0.1 M spermidine, and 8 μl of 2.5 M $CaCl_2$.
5. Keep vortexing for 5 min.
6. Allow the gold particles to settle for 10 min.
7. Pellet DNA-coated gold particles by brief centrifugation.
8. Remove the supernatant.
9. Add 80 μl of 100% ethanol and vortex briefly (*see* **Note 6**).
10. Centrifuge the tube briefly and remove the supernatant.
11. Add 80 μl of 100% ethanol and resuspend gold particles vigorously by pipetting (*see* **Note 7**).
12. Centrifuge the tube briefly and remove the supernatant.
13. Add 12 μl of 100% ethanol and resuspend gold particles by pipetting.

3.2 Particle Bombardment

The following procedures are carried out in a clean bench (*see* **Note 8**). For safe use of particle bombardment system, read the manual of PDS-1000/He (http://www.bio-rad.com/webroot/web/pdf/lsr/literature/M1652249.pdf).

1. Wipe the sample chamber of particle delivery system with 70% ethanol.
2. Sterilize a rupture disk (450 psi), a macrocarrier, a macrocarrier holder, and a stopping screen by dipping them into 100% ethanol and dry them out. Use tweezers for handling.
3. Place the macrocarrier inside the macrocarrier holder by using a macrocarrier insertion tool.
4. Load the 12-μl suspension of DNA-coated gold particles onto the macrocarrier-cum-macrocarrier holder. Allow time for drying.
5. Open the valve of a helium gas cylinder and set the pressure to over 650 psi.
6. Set the sterilized rupture disk, the stopping screen, and the macrocarrier-cum-macrocarrier holder with DNA-coated particles appropriately in the sample chamber.

7. Place duckweeds densely onto a 60-mm plastic dish without growth medium.
8. Set the dish on a target shelf 8 cm away from the stopping screen.
9. Close the sample chamber door and set the vacuum switch to the "VAC" position until the vacuum gauge reaches 26.5 mmHg, and then set the vacuum switch to the "HOLD" position.
10. Press and hold the "FIRE" switch until the rupture disk bursts with a small pop. Release the "FIRE" switch immediately after the disk ruptures.
11. Release the vacuum in the sample chamber by setting the vacuum switch to the "VENT" position.
12. After the vacuum is released, open the sample chamber door and remove the sample dish.
13. Add 8 mL of NF medium and 16 μl of 0.1 M D-luciferin (final concentration: 200 μM).
14. Leave the dish in a growth chamber for 1 day to fully express the transfected reporter gene.

3.3 Bioluminescent Imaging

In the following procedures, the EM-CCD camera, the motor-driven stage, the LED device are operated by using the HoKaWo imaging software.

1. Cool the EM-CCD to −80 °C.
2. Set EM gain to 0.
3. Open the lightproof box and set a 1-mm extension ring to the lens (F/0.95, focus-ring position/0.3).
4. Place the 60-mm dish with gene-transfected plants under the lens.
5. Observe the plants by live imaging and adjust focus by controlling the *Z*-axis motor.
6. Capture a bright-field image by snapshot (Fig. 2a).
7. Be sure to turn off the LED device and close the lightproof box door.
8. Wait 1 min for autofluorescence decay.
9. Set EM gain to 1200 and capture a bioluminescence image of a 5-min exposure (Fig. 2b).
10. Select a plant with a high density of luminescent spots.
11. Prepare a 35-mm plastic dish with a silicone rubber sheet on the bottom. Add 3.5 ml of NF medium and 7 μl of 0.1 M D-luciferin (final concentration: 200 μM). .
12. Transfer the selected plant to the 35-mm dish.

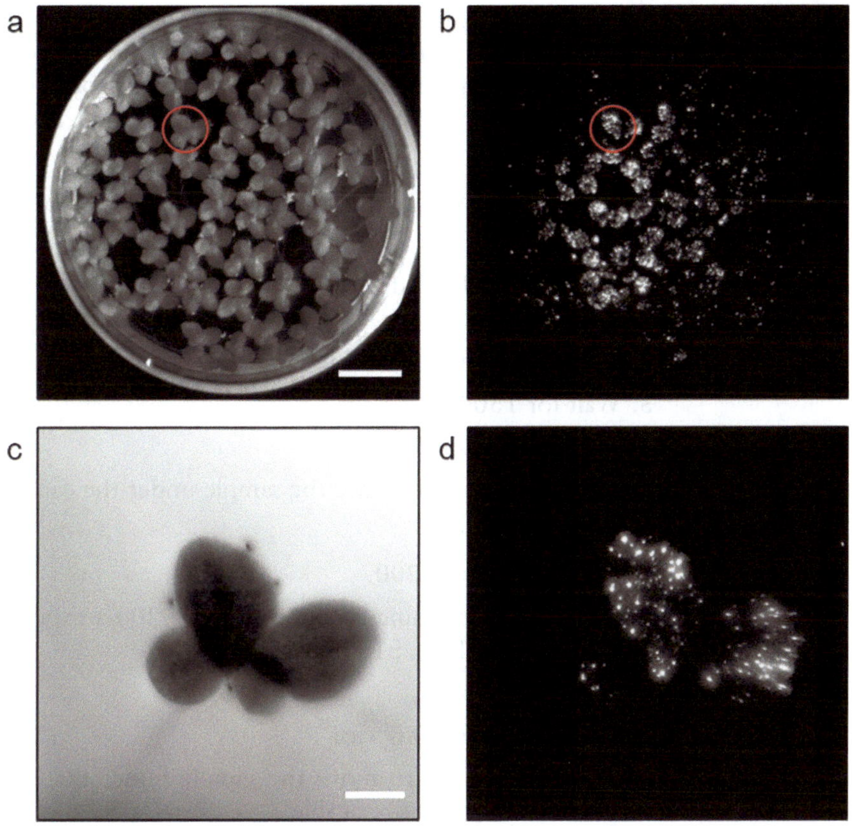

Fig. 2 Bright-field- and bioluminescent images of *Lemna gibba* transfected with *AtCCA1::LUC* by particle bombardment. A bright-field image (**a**) and a bioluminescent (**b**) image of duckweeds cultured in a 60-mm culture dish. The exposure time for the bioluminescent image is 300 s. The plant marked by a red circle was chosen for the following single-cell imaging. A bright-field image (**c**) and a bioluminescent (**d**) image for the single-cell imaging. The exposure time for the bioluminescent image was 200 s. Scale bars are 1 cm (**a**) and 2 mm (**b**)

13. Anchor the plant body with several insect pins fixed to the silicone rubber sheet.
14. Set a 16.5-mm extension ring assembly to the lens.
15. Place the 35-mm sample dish under the lens.
16. Set EM gain to 0.
17. Observe the plant by live imaging and adjust focus by controlling the *Z*-axis motor.
18. Capture a bright-field image by snap shot (Fig. 2c).
19. Be sure to turn off the LED device and close the lightproof box door.
20. Set EM gain to 1200.
21. Capture a bioluminescence image of a 60-s exposure to reconfirm the focus.

22. Set the exposure time to 200 s and start time-lapse imaging with 30 min interval (Fig. 2d). Before capturing each bioluminescence image, we allowed a 100-s dark period for autofluorescence decay. Prepare a script according to the following sequence.

An example of the operating sequence of time-lapse imaging (*see* **Note 9**).

1. Turn on the LED device.
2. Rotate the stage to place the sample under the illumination guided by optical fiber cable.
3. Wait for 1500 s.
4. Turn off the LED device.
5. Rotate the stage to place the sample under the camera lens.
6. Wait for 100 s.
7. Set EM gain to 1200.
8. Capture a bioluminescence image of a 200-s exposure (TIFF format file, 512 × 512 pixels).
9. Set EM gain to 0.
10. Turn on the LED device.
11. Rotate the stage to move the sample under the illumination guided by optical fiber cable.
12. Wait for 750 s before starting the next sequence (*see* **Note 10**).

3.4 Bioluminescence Quantification

The following image analysis is carried out with ImageJ (http://rsbweb.nih.gov/ij/).

1. Open time-series image files as a stack.
2. Increase image resolution by resizing the image size to four times. (Image>Adjust>Size, Interpolation: "None").
3. Correct position gaps of the sample between sequential images as follows (Fig. 3).
   (a) Select two series of corresponding luminescent spots in the stack of images (magenta and cyan in Fig. 3a).
   (b) Manually trace spot positions of a series (magenta in Fig. 3) in the stack of images.
   (c) Parallelly move images to set each spot position on the trace to the position in the first image.
   (d) Manually trace spot positions of another series (cyan in Fig. 3) in the stack of images after the parallel movement of (c).

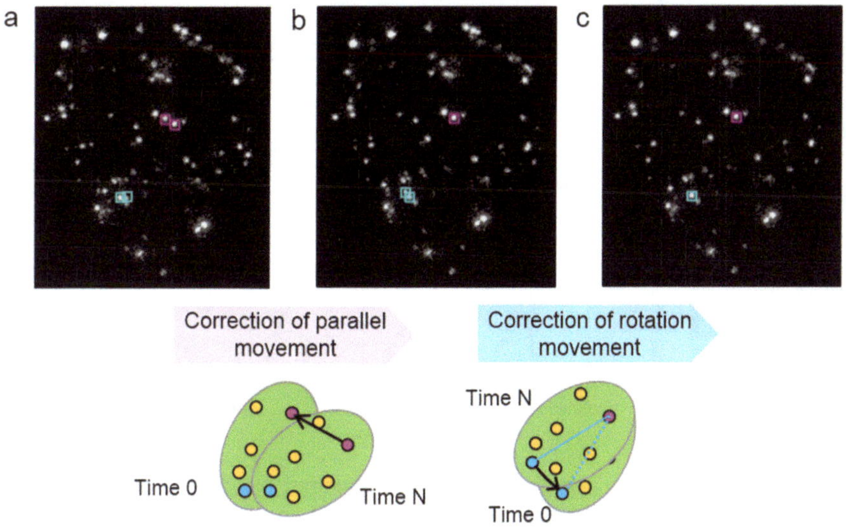

Fig. 3 Position correction of bioluminescent images of a sample drifting occasionally and slightly on the culture medium. Max projections generated with two images at different time points in the original stack (**a**), in the stack after the parallel movement correction (**b**) and in the stack after the rotation movement correction of (**b**, **c**). For the parallel and rotation movement correction, positions of luminous spots marked magenta and cyan squares were manually traced, respectively

(e) Rotate images around the point of the parallel movement correction (magenta in Fig. 3b) to set each spot position on the trace to the position in the first image.

4. Manually set a region of interest (ROI) for each luminescent spot in the stack of images after the parallel and rotation movement (*see* **Note 11**).
5. Measure the signal intensity of each luminescent spot in individual images as the integrated density of the ROI (*see* **Note 12**).
6. Calculate the background signal of pixels in each image as the median of signal intensities within the ROI (100×100 pixels) that is set at the area showing no luminescent spots (*see* **Note 13**).
7. Quantify the luminescent intensity of each luminescent spot by the following equation (Fig. 4) (*see* **Note 14**).

   Number of photons = (Signal intensity − Background intensity) × Conversion factor/(Analog gain × EM gain × Conversion efficiency).

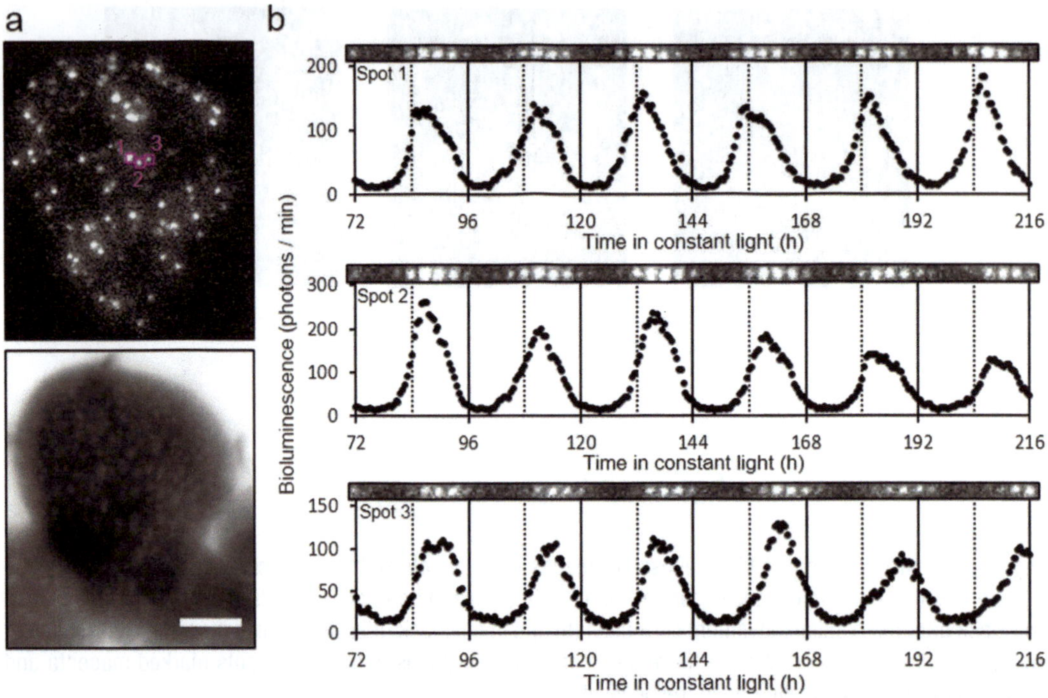

Fig. 4 Bioluminescent circadian rhythms of individual cells in a plant under constant light conditions. (**a**) Bioluminescence (*top*) and bright-field (*bottom*) images of a frond (leaf like structure) of *L. gibba* transfected with *AtCCA1::LUC*. Bioluminescent traces (30-min interval) of the three spots shown in (**a**) are represented in (**b**). Scale bar is 1 mm. Sequential images with 3 h intervals are shown above each trace

4 Notes

1. The transient gene transfection by particle bombardment can be applied to any duckweed species as well [20].
2. White light was supplied by fluorescent lamps (FLR40SEX-W/M/36-HG; NEC). *L. gibba* plants are aseptically grown on 60-ml medium in a sterilized 200-ml Erlenmeyer flask plugged with cotton. New stock cultures are made every week. Growth conditions can be changed according to experimental designs [16, 20].
3. The detail of *AtCCA1:LUC* is described in Nakamichi et al. [21]. Plasmid DNA is purified by Plasmid Midi kit (QIAGEN). Plasmid DNA purified by a mini-preparation procedure is also available. Reporter constructs for Agrobacterium-mediated transformation are not recommended due to less transfection efficiencies.

4. Use a motor unit with a magnetic position sensor. That with an optical position sensor increases the background signals in bioluminescent imaging.
5. Before pipetting, 1-min sonication using an ultrasonic washing machine helps a uniform suspension of gold particles.
6. The gold particle pellet becomes stiff if the DNA coating is successful.
7. A uniform suspension of gold particles appears to be important for a high efficiency of gene transfection.
8. Aseptic techniques are required for long-term monitoring.
9. This example is for a measurement under constant light conditions. For measurement under dark conditions, skip the command for "Turn on the LED device".
10. Make fine adjustments of waiting time to make the loop period precise.
11. We set a 6×6 pixel ROI (24×24 pixels in the resized image) for a spot. One pixel size is ~25 μm in our imaging system. Thus, ~150 μm is a limit resolution of this imaging technique.
12. Luminescence intensities of spots in the same plant show a log-normal distribution of a large range (ca. 1000-fold differences between minimum and maximum) [15]. Our previous observation suggested that this large variation was partially due to the transient transfection by particle bombardment; the luminescence intensity was affected by the intracellular end-point of introduced gold particles and the amount of coated DNA [22]. Thus, the differences found in bioluminescence intensities among cells do not always reflect differences of transcriptional activities for the reporter construct among them.
13. There are unavoidable background signals such as the thermal noise of CCD sensor, delayed fluorescence of plants, and the plastic dish. Therefore, when light conditions are changed during the monitoring, background signals also change.
14. In our conditions, we calculate as follows: Number of photons = (Signal intensity − Background intensity) × 5.8/(1 × 1200 × 0.9).

References

1. Millar AJ, Short SR, Chua NH, Kay SA (1992) A novel circadian phenotype based on firefly luciferase expression in transgenic plants. Plant Cell 4:1075–1087
2. Millar AJ, Short SR, Hiratsuka K et al (1992) Firefly luciferase as a reporter of regulated gene expression in higher plants. Plant Mol Biol Report 10:324–337
3. Kondo T, Strayer CA, Kulkarni RD et al (1993) Circadian rhythms in prokaryotes: luciferase as a reporter of circadian gene expression in cyanobacteria. Proc Natl Acad Sci U S A 90:5672–5676

4. Brandes C, Plautz JD, Stanewsky R et al (1996) Novel features of drosophila period transcription revealed by real-time luciferase reporting. Neuron 16:687–692
5. Geusz ME, Fletcher C, Block GD et al (1997) Long-term monitoring of circadian rhythms in c-*fos* gene expression from suprachiasmatic nucleus cultures. Curr Biol 7:758–766
6. Jolma IW, Laerum OD, Lillo C, Ruoff P (2010) Circadian oscillators in eukaryotes. Wiley Interdiscip Rev Syst Biol Med 2:533–549
7. Yamaguchi S, Isejima H, Matsuo T et al (2003) Synchronization of cellular clocks in the suprachiasmatic nucleus. Science 302:1408–1412
8. Buhr ED, Yoo S-H, Takahashi JS (2010) Temperature as a universal resetting cue for mammalian circadian oscillators. Science 330:379–385
9. Leise TL, Wang CW, Gitis PJ, Welsh DK (2012) Persistent cell-autonomous circadian oscillations in fibroblasts revealed by six-week single-cell imaging of PER2::LUC bioluminescence. PLoS One 7:e33334
10. Ono D, Honma S, Honma K (2016) Differential roles of AVP and VIP signaling in the postnatal changes of neural networks for coherent circadian rhythms in the SCN. Sci Adv 2: e1600960
11. Millar AJ, Carré IA, Strayer CA et al (1995) Circadian clock mutants in *Arabidopsis* identified by luciferase imaging. Science 267:1161–1163
12. Nakamichi N, Kita M, Ito S et al (2005) Pseudo-response regulators, PRR9, PRR7 and PRR5, together play essential roles close to the circadian clock of *Arabidopsis thaliana*. Plant Cell Physiol 46:686–698
13. Fukuda H, Nakamichi N, Hisatsune M et al (2007) Synchronization of plant circadian oscillators with a phase delay effect of the vein network. Phys Rev Lett 99:098102
14. Wenden B, Toner DLK, Hodge SK et al (2012) Spontaneous spatiotemporal waves of gene expression from biological clocks in the leaf. Proc Natl Acad Sci U S A 109:6757–6762
15. Muranaka T, Kubota S, Oyama T (2013) A single-cell bioluminescence imaging system for monitoring cellular gene expression in a plant body. Plant Cell Physiol 54:2085–2093
16. Hillman WS (1961) The Lemnaceae, or duckweeds: a review of the descriptive and experimental literature. Bot Rev 27:221–287
17. Wang W, Haberer G, Gundlach H et al (2014) The *Spirodela polyrhiza* genome reveals insights into its neotenous reduction fast growth and aquatic lifestyle. Nat Commun 5:1–13
18. Muranaka T, Oyama T (2016) Heterogeneity of cellular circadian clocks in intact plants and its correction under light-dark cycles. Sci Adv 2:e1600500
19. Okada M, Muranaka T, Ito S, Oyama T (2017) Synchrony of plant cellular circadian clocks with heterogeneous properties under light/dark cycles. Sci Rep 7:317
20. Muranaka T, Okada M, Yomo J et al (2015) Characterisation of circadian rhythms of various duckweeds. Plant Biol 17:66–74
21. Nakamichi N, Ito S, Oyama T et al (2004) Characterization of plant circadian rhythms by employing *Arabidopsis* cultured cells with bioluminescence reporters. Plant Cell Physiol 45:57–67
22. Yamashita T, Iida A, Morikawa H (1991) Evidence that more than 90% of b-glucuronidase-expressing cells after particle bombardment directly receive the foreign gene in their nucleus. Plant Physiol 97:829–831

Index

A

Adeno-associated vectors 161
Alpha-synuclein 147–154, 157
AmBisome 83, 91, 93, 94, 101–104
Aminoluciferin (aLuc) 219–227
Androgen 30, 31, 36, 38
Anesthesia 72, 73, 83, 89, 90, 95, 105, 114, 131, 134, 135, 141, 154, 155, 157, 193–195
Annexin V 15–18, 20–26
Antibody 150, 153, 156, 157, 178
Antifungal 127–142
Apoptotic 15–26
Arabidopsis 231, 232
Astrocytes 148, 150
Astrogliosis 148, 157
Autobioluminescence 55, 191–200

B

17β-estradiol 31
BALB/c mouse 81–105
Base of the tail 82, 89–92, 94, 97, 100, 102–104
Bioassays 29–40
Bioavailability 30, 101
Biodistribution 64, 141, 191
Biofilm 108, 109, 111–114, 117, 119, 121, 124, 127–142
Bioimaging 55, 162
Bioluminescence 3–13, 15–26, 29–40, 43–51, 53–64, 69–79, 81–105, 109, 111–114, 116, 117, 120–122, 124, 127–142, 147–158, 161–173, 177–192, 199, 203–209, 211–216, 219–227, 231–241
Bioluminescence resonance energy transfer (BRET) 15–26
Bioluminescent imaging (BLI) 9, 43–51, 53–64, 69–79, 161–173, 177–191, 204, 219, 224, 231–241
Biophotonic 108, 109, 112, 119
Bioreporters 30, 31, 37
Brain 148, 150, 151, 155–158, 172, 182, 220, 221, 225, 226
Breakthrough 57–61

C

Cancer 15, 29, 203, 204, 208, 209, 211–213, 215
Candida 127–142
Central nervous system (CNS) 148, 150, 151, 157
Central venous catheter (CVC) 128, 129
Cerebral palsy 162
Chemotaxis 59
Chick embryo 212
Chorioallantoic membrane 211–216
Chromosome 30
Circadian rhythms 231–241
Coelenterazine (CTZ) 16, 18, 22–25, 110, 117, 118, 120, 124, 129, 131, 133, 135, 139–141, 205, 207
Commensal 107–109
Confocal laser scanning microscope 153, 156
Conscious imaging 161–173
CRISPR/Cas9 232
Cutaneous leishmaniasis (CL) 81–105

D

Dental caries 107
Dental plaque 109
Dihydrotestosterone (DHT) 31, 34–40
D-luciferin (dLuc) 5, 8, 9, 11–13, 110, 113, 117, 120, 124, 132, 133, 135, 141, 152, 155, 158, 165, 172, 173, 179, 181, 191, 212, 213, 215, 219, 220, 225, 226, 233, 236
Drug discovery 3, 81–105
Drug efficacy 91–94, 102–104
Drug screening 82
Duckweed 232, 236, 237, 239

E

Ear 82, 90, 92, 94–97, 100, 101, 103
Electron multiplying charge-coupled device (EMCCD) 6, 9, 44, 46
Elution 17, 57, 60, 61, 170
Emerging contaminants 29
Endocrine disrupting compound (EDC) 29–40

Steven Ripp (ed.), *Bioluminescent Imaging: Methods and Protocols*, Methods in Molecular Biology, vol. 2081, https://doi.org/10.1007/978-1-4939-9940-8, © Springer Science+Business Media, LLC, part of Springer Nature 2020

Endocrinology ... 29
Enteric redmouth disease (ERM) ... 69, 70, 72
Escherichia coli ... 18, 44, 46, 47, 49, 50, 70–72, 75, 77, 79
Estradiol ... 31
Estrogen ... 30, 31, 34, 36–39
Extracellular vesicles (EVs) ... 203–209

F

Fetal bovine serum (FBS) ... 5, 82, 83, 96, 130, 133, 134, 163, 164, 167–169, 192–194, 204, 215
Filoviridae ... 177
Fish ... 69–79, 212
Flow cytometry ... 23, 24
Fluorescence ... 15, 16, 19, 20, 23–26, 43–46, 48, 49, 54, 55, 138, 212, 221
Footpad ... 81–83, 89, 92, 94, 97, 100, 103, 104
Fungus ... 128
Fusion tags ... 50
Fusobacterium nucleatum ... 110, 112, 117, 124

G

G418 ... 192, 193, 199
Glycosylphosphatidylinositol (GPI) ... 128
Golden Syrian Hamster (GSH) ... 81, 82

H

Half maximal effective concentration (EC_{50}) ... 39
Haptotaxis ... 59
Hemocytometer ... 82, 85, 86, 98, 193, 194
High performance liquid chromatography (HPLC) ... 31–33, 40, 169, 170
High-throughput ... 3–13, 29–40
Histology ... 148, 212
Human embryonic kidney (HEK293) ... 4, 5, 7–11, 192, 193, 199
Humic acid (HA) ... 55, 56, 62

I

Ilux ... 43–51
Imaging modalities
  ex ovo ... 211
  ex vivo ... v, 179, 182, 188, 192
  in ovo ... 211
  in situ ... 53, 54
  in vitro ... 111, 113, 128, 129, 131–133, 139, 140, 192, 220, 223
  in vivo ... 43, 55, 69–79, 81–105, 108, 111, 112, 114, 119, 127–142, 165, 166, 177–200, 203–209, 213, 224, 226, 231
Immunocompromised ... 128, 192, 193
Immunofluorescence ... 148, 150, 152, 153, 156
Implant ... 130, 134, 224
Infections ... 43, 69, 70, 72–74, 79, 81–86, 88–100, 102–104, 108, 109, 112, 114, 116–119, 121, 124, 127–129, 131, 135, 177, 178, 181, 182, 184, 185, 187, 190
Inflammation ... 162
Inhibitory concentration (IC_{20}) ... 35, 37
Injection routes
  intradermal ... 81, 89, 90, 141
  intraglossal ... 148, 149, 151, 154, 157
  intramuscular ... 118
  intraorganellar ... 193
  intraperitoneal (IP) ... 72–74, 78, 79, 87, 88, 91, 117, 118, 141, 148, 149, 151, 154, 157, 158, 172, 181, 191, 193, 225, 226
  intravenous (IV) ... 101, 130, 136, 139, 141, 172, 189, 191, 208
  subcutaneous (SC) ... 83, 89, 101, 118, 128, 134–136, 141, 191, 193, 199
Internal ribosomal entry site (IRES) ... 204
In vivo Imaging System (IVIS) ... 55, 56, 59, 62–64, 70, 73, 75, 77–79, 83, 86–88, 91, 97, 100, 109, 111, 114, 117, 131, 139, 152, 166, 172, 179, 181, 182, 193, 194, 205, 207, 208, 213, 215, 221, 223–225
In-Vivo Xtreme ... 109, 111, 114, 116
Ion Chromatography ... 58
Isoflurane ... 83, 87, 89, 90, 97, 100, 111, 114, 117, 135, 141, 152, 154, 155, 162, 166, 172, 179, 189, 193–195, 200, 205, 207, 224, 225

K

Ketamine ... 83, 89, 94, 97, 131, 134, 141, 152, 154

L

Laser induced breakdown spectroscopy (LIBS) ... 54
Leishmania major (*L. major*) ... 82, 84, 89–92, 101–104
Lemna gibba ... 232, 233, 237, 239, 240
Lentivirus ... 161, 165, 167, 168, 172, 173, 204
Lesion ... 81–84, 89–91, 93–95, 101–105
Lewy body (LB) ... 17, 20, 45, 47, 147, 163, 166
Ligation ... 75, 77
Lipids ... 203

Longitudinal 82, 192
Luciferase
  bacterial 108, 117, 120
  click beetle 112, 119, 120
  Cypridina 109, 112, 119, 120
  firefly 82, 109, 119, 120, 128, 148, 153, 178, 191, 192, 212, 213, 215, 219
  Gaussia 128, 129, 137, 138
  Luciola red 109, 110, 112, 113, 116, 119, 120, 122
  Renilla 16, 17, 109, 119, 120, 124, 204
  synthetic 108, 112, 128, 192
Luciferin 43, 83, 84, 86–88, 96, 99, 100, 129, 165, 173, 189, 191, 192, 200, 219, 220
Luminescence 9, 15, 16, 18, 22, 24, 25, 30, 59, 62, 77, 78, 82, 86, 88, 99, 101, 111, 121, 131, 139, 180, 189, 232, 241
Luminometer 4, 6, 9, 30, 111, 139, 179, 180
luxCDABE 30, 44, 69, 70, 74, 192
Lyophilization 12

M

Marburg virus 177–190
Metacyclic 98
Metastasis 93, 103
Microbiome 107–125
Microgliosis 148, 150, 157
Microscope 156
Microscopy 4, 6, 9, 16, 18, 20, 44–49, 85, 96, 138, 152, 153, 212
Microtiter plate 30, 34, 35, 56, 59, 71, 77, 205, 207
Mold 127
Mouse 81–105, 111, 114, 116, 117, 119, 121, 122, 129, 137, 139, 147–158, 162, 178, 181, 187, 189, 191–200, 205, 207, 224–226
Multiplex imaging 107–125
Multispectral imaging (MSI) 54
Murine 83, 97, 100, 101, 107–125, 178, 189, 211, 212
Muscle 148, 182
Mutagenesis 44, 113
Mycology 84

N

N-cyclobutylaminoluciferin (cybLuc) 220
Near-infrared (NIR) 16, 18, 24–26
Neurodegenerative disease 15
Neuroinflammation 147–158
Neuroinvasion 148, 157
Noninvasive 55, 82, 178, 192

O

Oral 107–125
Overlap extension PCR (OE-PCR) 110, 113, 114

P

Parasite 81–105
Parkinson's disease (PD) 147–158
Particle bombardment 232, 233, 235, 237, 239, 241
Pathogen 53, 54, 128, 132, 178, 221
Pathology 81, 108, 147, 148, 157
Periodontitis 107–109
Peripheral inoculation 152
Photorhabdus luminescens 30, 44, 69, 70, 192
Pixel 44, 46, 48, 54, 87, 101, 187, 234, 238, 239, 241
Plasmid 5, 7–9, 11, 12, 17, 18, 20, 45, 47, 72–78, 129, 130, 162–168, 179, 235, 239
Polymerase chain reaction (PCR) 17, 20, 74, 76, 77, 108, 110, 114, 153, 171
Polymicrobial 107–125
Prevotella nigrescens 110, 112, 117, 118, 124
Prion 147, 148
Profile distribution 61
Promastigotes 83–86, 89–92, 95, 96, 98, 99, 101
Pseudomonas fluorescens 55, 56
Pseudotyping 162, 178–180, 185
Pseudovirus 177–180

Q

Quantum dot (QD) 16, 19, 20, 102, 103

R

Rabbit 128
Radiance 148, 151, 155, 185, 190
Radioisotope 15
Rainbow trout 70, 72–74
Rat 128–131, 134, 135, 139, 141
Real-time 53, 55, 56, 60, 69, 153, 178, 191–200
Region of interest (ROI) 9, 62, 64, 74, 75, 77, 78, 88, 89, 101, 133, 134, 136, 155, 187, 188, 190, 197, 198, 200, 225, 239, 241
Relative light unit (RLU) 36, 38, 74, 75, 78, 99, 113, 119, 121

Renilla reniformis 120, 204
Reporter gene 3–13, 82, 161, 162, 177, 236
Rodent 86, 87, 97, 99, 108, 111, 114, 119, 152, 161, 177–190, 212, 219–227

S

Saccharomyces cerevisiae 30, 31
Salivary gland 81, 95, 96, 105
Sand 55, 56, 60–63
Sand fly 81, 94–96
Scaffold 211
Single cell imaging 47, 231–241
Soil 53–64
Somatotransgenic 162
Spectral imaging 54
Spectrometers 16, 18, 22, 25, 220
Spheroid 3–13
Stem cells 211
Streptococcus 69, 109
Surface-enhanced Raman spectroscopy (SERS) 54
Surveillance 192
Symbiosis 107, 108
Synucleinopathies 147

T

Terahertz time-domain spectroscopy (THz-TDS) 54
Three-dimensional (3D) 3–13, 64, 103
Tobacco 231
Tongue 107, 148, 154, 157
Tracking .... v, 53–64, 69, 74, 88, 148, 203–209
Transcriptional terminator 74–76
Transfection 5, 7–12, 163, 164, 167–169, 173, 179, 180, 185, 199, 205, 232, 236, 237, 239–241
Transformation 77–79, 239
Transgenics 129, 139, 161, 162, 221, 225, 231, 232
Tumor 43, 203–209, 211, 212, 214, 215, 219, 220, 224, 225

V

Vargulin 120
Vector 5, 10, 17, 20, 47, 50, 70, 72, 73, 75, 81, 161, 162, 165–173, 178, 204, 205, 215

W

Wastewater 30

X

Xenograft 213–215, 224, 225

Y

Yeast 29–40, 45, 55, 110, 127–130, 163
Yersinia ruckeri 69–79

MIX
Papier aus verantwortungsvollen Quellen
Paper from responsible sources
FSC® C105338

If you have any concerns about our products,
you can contact us on
ProductSafety@springernature.com

In case Publisher is established outside the EU,
the EU authorized representative is:
Springer Nature Customer Service Center GmbH
Europaplatz 3, 69115 Heidelberg, Germany

Printed by Libri Plureos GmbH
in Hamburg, Germany